T0269490

SpringerBriefs in Earth Sciences

SpringerBriefs in Earth Sciences present concise summaries of cutting-edge research and practical applications in all research areas across earth sciences. It publishes peer-reviewed monographs under the editorial supervision of an international advisory board with the aim to publish 8 to 12 weeks after acceptance. Featuring compact volumes of 50 to 125 pages (approx. 20,000–70,000 words), the series covers a range of content from professional to academic such as:

- timely reports of state-of-the art analytical techniques
- bridges between new research results
- snapshots of hot and/or emerging topics
- literature reviews
- in-depth case studies

Briefs will be published as part of Springer's eBook collection, with millions of users worldwide. In addition, Briefs will be available for individual print and electronic purchase. Briefs are characterized by fast, global electronic dissemination, standard publishing contracts, easy-to-use manuscript preparation and formatting guidelines, and expedited production schedules.

Both solicited and unsolicited manuscripts are considered for publication in this series.

More information about this series at http://www.springer.com/series/8897

Alik Ismail-Zadeh • Alexander Korotkii
Igor Tsepelev

Data-Driven Numerical Modelling in Geodynamics: Methods and Applications

 Springer

Alik Ismail-Zadeh
Institute of Applied Geosciences
Karlsruhe Institute of Technology
Karlsruhe, Germany

Institute of Earthquake Prediction Theory
and Mathematical Geophysics
Russian Academy of Sciences
Moscow, Russia

Igor Tsepelev
Institute of Mathematics and Mechanics
Russian Academy of Sciences
Yekaterinburg, Russia

Alexander Korotkii
Institute of Mathematics and Mechanics
Russian Academy of Sciences
Yekaterinburg, Russia

ISSN 2191-5369   ISSN 2191-5377   (electronic)
SpringerBriefs in Earth Sciences
ISBN 978-3-319-27800-1   ISBN 978-3-319-27801-8   (eBook)
DOI 10.1007/978-3-319-27801-8

Library of Congress Control Number: 2016934598

Printed on acid-free paper

This Springer imprint is published by Springer Nature
The registered company is Springer International Publishing AG Switzerland

Preface

La physique ne nous donne pas seulement l'occasion de résoudre des problems, . . . elle nous fait pressentir la solution (Physics gives us not only the opportunity to solve problems, . . . it helps us to anticipate the solution)

(Henri Poincaré, 1854–1912)

Dynamic processes in the Earth's interior and on its surface can be described by geodynamic models. These models can be presented by a mathematical problem comprising a set of partial differential equations with relevant conditions at the model boundary and at the initial time. The mathematical problem can be then solved numerically to obtain future states of the model. Meanwhile the initial conditions in the geological past or some boundary conditions at the present are unknown, and the question of how to "find" the conditions with a sufficient accuracy attracts attention in the field of geodynamics. One of the mathematical approaches is data assimilation or the use of available data to reconstruct the initial state in the past or boundary conditions and then to model numerically the dynamics of the Earth starting from the reconstructed conditions.

Quantitative geodynamic models have been developed with the advent of powerful computers. Since the 1980s simple data assimilation methods began to be employed in geodynamic modelling, and since the 2000s more sophisticated methods have been in use. The assimilation models have shown a capability to reconstruct thermal and dynamic characteristics of the solid Earth in the geological past. However, the models require accurate geophysical, geodetic, geochemical, and geological data. The observations, although growing in size, are still sparse and irregularly distributed in space and time. Therefore, efficient methods for data assimilation are needed for accurate reconstructions and modelling.

In this book we describe the methods and numerical approaches for data assimilation in geodynamical models and present several applications of the described methodology to relevant case studies. The book starts with a brief overview of the basic principles in data-driven geodynamic modelling, inverse problems, and data assimilation methods, which is then followed by methodological chapters on backward advection, variational (or adjoint), and quasi-reversibility methods. The chapters are accompanied by those presenting applications of the methods to solving some geodynamic problems such as mantle plume evolution, lithosphere dynamics, salt diapirism, and a volcanic lava flow. These applications might present an interest to the hydrocarbon industry and to experts dealing with geohazards and risk mitigation. For example, the knowledge of sedimentary basin evolution complicated by deformations due to salt tectonics can help in oil and gas exploration; the understanding of the stress-strain evolution and stress localisation can provide an insight on the preparation of large earthquakes; volcanic lava flow assessments can mitigate a potential risk for population and infrastructure.

We have to apologise that the book does not contain all methods for data assimilation, but only frequently used in geodynamic modelling. However, we believe that the methods and the applications described here will be helpful for understanding how geo-data can be utilised to resolve quantitatively some problems in geodynamics.

Karlsruhe, Germany   Alik Ismail-Zadeh
Yekaterinburg, Russia   Alexander Korotkii
Yekaterinburg, Russia   Igor Tsepelev

Acknowledgements

This book would have not been written, if two groups of experts, one from the Institute of Earthquake Prediction Theory and Mathematical Geophysics in Moscow (IEPT) and another from the Institute of Mathematics and Mechanics (IMM) in Yekaterinburg (both of the Russian Academy of Sciences), did not meet in the early 1990s under the umbrella of a joint project sponsored by the International Science and Technology Centre. The aim of the project was to convert pure mathematics to application in geophysics. The project started with the development of new software for geodynamic modelling and its application to the development of sedimentary basins. Later the cooperation continued in the area of thermal convective flows in the mantle and lithosphere dynamics. Our recent collaboration is related to volcanic lava flows sponsored by the Russian Science Foundation. All studied topics have been associated with inverse geodynamic problems and data-driven numerical modelling or data assimilation in geodynamics.

We thank our colleagues for fruitful discussions on data assimilation and numerical modelling in geodynamics: Hans-Peter Bunge, Claude Boucher, Andreas Fichtner, Alessandro Forte, Alexandre Fournier, Taras Gerya, Michael Ghil, Michael Gurnis, Satoru Honda, Dmitry Kovtunov, Oleg Melnik, Boris Naimark, Michael Navon, Yuri Podladchikov, Gerald Schubert, Alexander Soloviev, Bernhard Steinberger, Paul Tackley, Olivier Talagrand, Chris Talbot, Valery Trubitsyn, Vladimir Vasin, Yuri Volozh, and David Yuen. We acknowledge with great pleasure the support from the Institute of Applied Geosciences, Karlsruhe Institute of Technology (Karlsruhe, Germany), and from the Institute of Earthquake Prediction Theory and Mathematical Geophysics (Moscow) and the Institute of Mathematics and Mechanics (Yekaterinburg), Russian Academy of Sciences (Russia).

Contents

Chapter 1
Introduction

Abstract In this chapter, we introduce direct and inverse problems as well as well- and ill-posed problems, which are characterised by the existence, uniqueness, and stability of the problem solution. We present some examples of unstable problems and discuss the basic elements in forward and backward numerical modelling and the errors associated with the modelling. Finally we briefly review the methods for data assimilation used in geodynamic modelling.

Keywords Data assimilation • Methods • Numerical modelling • Well-posedness • Ill-posedness • Solution stability • Errors

Geodynamics, whose past and current behaviours are of great scientific interest, deals with dynamic processes in the Earth's interior and on its surface. Mantle convection, hotspots and mantle plumes, lithosphere dynamics and plate subduction as well as their surface manifestation as volcanism, seismicity, and sedimentary basins evolution are among principal geodynamic problems (e.g. Turcotte and Schubert 2002). Many geodynamic problems can be described by mathematical models, i.e. by a set of partial differential equations and boundary and/or initial conditions defined in a specific domain. A mathematical model links the causal characteristics of a geodynamic process with its effects. The causal characteristics of the model process include, for example, parameters of the initial and boundary conditions, coefficients of the differential equations, and geometrical parameters of a model domain. The aim of the *direct* mathematical problem is to determine the effects of a geodynamic model process based on the knowledge of its causes, and hence to find a solution to the mathematical problem for a given set of parameters and coefficients. An *inverse* problem is the opposite of a direct problem. An inverse problem is considered when there is a lack of information on the causal characteristics but information on the effects of the geodynamic process exists.

In this book we refer to a data-driven numerical model in geodynamics, when a numerical model derived from the mathematical model describing a geodynamic process is considered in the cases of known effects of the geodynamic process (available geophysical, geological, geochemical, geodetic, and other data) but (some) unknown causal characteristics of the model. The goal of data-driven

© The Author(s) 2016
A. Ismail-Zadeh et al., *Data-Driven Numerical Modelling in Geodynamics: Methods and Applications*, SpringerBriefs in Earth Sciences,
DOI 10.1007/978-3-319-27801-8_1

numerical modelling is to determine the model characteristics. Such approach is known also as data assimilation. There are two approaches in data assimilation. The classical approach considers a mathematical model as a true model and available geo-data as a true data set with some measurement errors, and the goal is to recover the true model (e.g., initial or boundary conditions). Another way to treat a mathematical model is the Bayesian approach, where the model is considered as a random variable, and the solution is a probability distribution for the model parameters (Aster et al. 2005). Here we consider the classical data assimilation approach and use the term "data assimilation in geodynamic modelling" assuming that it is as a synonym to "data-driven numerical modelling in geodynamics".

1.1   Inverse Problems in Geodynamics

Inverse problems can be subdivided into time-reverse or retrospective problems (e.g. to restore the development of a geodynamic process), coefficient problems (e.g. to determine the coefficients of the model equations and/or boundary conditions), geometrical problems (e.g. to determine the location of heat sources in a model domain or the geometry of the model boundary), and some others. Inverse problems are often ill-posed. The idea of well- (and ill-) posedness in the theory of partial differential equations was introduced by Hadamard (1902). A mathematical model of a geodynamic problem is considered to be well-posed if (i) a solution to this problem exists, and the solution is (ii) unique and (iii) stable. Problems for which at least one of these three properties does not hold are called *ill-posed*. The requirement of stability is the most important one. If a problem lacks the property of stability then its solution is almost impossible to compute, because computations are polluted by unavoidable errors. If the solution of a problem does not depend continuously on the initial data, then, in general, the computed solution may have nothing to do with the true solution.

   For example, the retrospective (inverse) problem of thermal convection in the mantle is an ill-posed problem, since the backward heat problem, describing both heat advection and conduction through the mantle backwards in time, possesses the properties of ill-posedness or instability (Kirsch 1996). In particular, the solution to the problem does not depend continuously on the initial data. This means that small changes in the present-day temperature field may result in large changes of predicted mantle temperatures in the past. Let us explain this statement using two simple problems related to the one-dimensional (1-D) diffusion equation and the two-dimensional (2-D) Laplace equation.

Example 1.1 Consider the following problem for the 1-D backward diffusion equation:

$$\partial u\,(t, x)\,/\partial t = \partial^2 u\,(t, x)\,/\partial x^2, \quad 0 \le x \le \pi, \quad t < 0$$

with the following boundary and initial conditions

$$u\,(t,0) = 0 = u\,(t,\pi)\,,\quad t \le 0,\quad u\,(0,x) = \phi_n(x),\quad 0 \le x \le \pi.$$

At the initial time the function $\phi_n(x)$ is assumed to take the following two forms:

$$\phi_n(x) = \frac{\sin((4n+1)\,x)}{4n+1}\quad \text{and}\quad \phi_0(x) \equiv 0.$$

We note that

$$\max_{0 \le x \le \pi} |\phi_n(x) - \phi_0(x)| \le \frac{1}{4n+1} \to 0 \text{ at } n \to \infty. \tag{1.1}$$

The two solutions of the problem

$$u_n\,(t,x) = \frac{\sin((4n+1)x)}{4n+1}\,\exp\left(-(4n+1)^2 t\right)\quad \text{at}\quad \phi_n(x) = \phi_n \text{ and}$$

$$u_0\,(t,x) \equiv 0 \quad \text{at}\quad \phi_n(x) = \phi_0$$

correspond to the two chosen functions of $\phi_n(x)$, respectively. At $t = -1$ and $x = \pi/2$

$$u_n\,(-1,\pi/2) = \frac{1}{4n+1}\,\exp\left((4n+1)^2\right)\quad \text{at}\quad n \to \infty. \tag{1.2}$$

At large n two closely set initial functions ϕ_n and ϕ_0 are associated with the two strongly different solutions at $t = -1$ and $x = \pi/2$. Hence, a small error in the initial data (1.1) can result in very large errors in the solution to the backward problem (1.2), and therefore the solution is unstable, and the problem is ill-posed in the sense of Hadamard.

Example 1.2 Now we consider a problem for the 2-D Laplace equation. This problem can be formulated as:

$$\partial^2 u\,(y,x)\,/\partial y^2 = -\partial^2 u\,(y,x)\,/\partial x^2,\quad y \le 0,$$

with the following initial conditions

$$u\,(0,x) = 0,\quad \partial u\,(y,x)\,/\partial y|_{y=0} = n^{-1}\sin(nx).$$

The solution of the problem is

$$u_n\,(y,x) = n^{-1}\,(e^{ny} + e^{-ny})\sin(nx). \tag{1.3}$$

Note that $\max \left| u_y \left(0, x \right) \right| = \max \left| n^{-1} \sin(nx) \right| \leq n^{-1} \to 0$ at $n \to \infty$. Meanwhile the solution (1.3) $u_n \to \infty$ at $n \to \infty$ for $x \neq \pi n$, $n = 0, \pm 1, \pm 2, \dots$, and therefore the solution is unstable, and the problem is ill-posed in the sense of Hadamard.

Despite the fact that many inverse problems are ill-posed, there are some methods for solving them. The idea of conditionally well-posed problems and the regularization method were introduced by Tikhonov (1963). According to Tikhonov, a class of admissible solutions to conditionally ill-posed problems should be selected to satisfy the following conditions: (i) a solution exists in this class, (ii) the solution is unique in the same class, and (iii) the solution depends continuously on the input data (that is, the solution is stable). The Tikhonov regularization is essentially a trade-off between fitting the observations and reducing a norm of the solution to the mathematical model of a geodynamic problem. Using two examples we show below the differences between the Hadamard's and Tikhonov's approaches to ill-posed problems.

Example 1.3 Consider the problem for the 1-D backward diffusion equation (similar to the problem presented in Example 1.1):

$$\partial u \left(t, x \right) / \partial t = \partial^2 u \left(t, x \right) / \partial x^2, \quad 0 \leq x \leq \pi, \quad -T \leq t < 0,$$

with the boundary and initial conditions

$$u \left(t, 0 \right) = 0 = u \left(t, \pi \right), \quad t \leq 0, \; u \left(0, x \right) = \phi_0(x), \quad 0 \leq x \leq \pi.$$

The solution of the problem satisfies the inequality:

$$\| u \left(\cdot, t \right) \| \leq \| u \left(\cdot, T \right) \|^{-t/T} \| u \left(x, 0 \right) \|^{1+t/T},$$

where the norm is presented as $\| u \left(\cdot, t \right) \|^2 \equiv \int_0^\pi u^2 \left(x, t \right) dx$. We note that the inequality

$$\| u \left(\cdot, t \right) \| \leq M^{-t/T} \| u_0 \|^{1+t/T}. \tag{1.4}$$

is valid in the class of functions $\| u \left(\cdot, t \right) \| \leq M = const$ (Samarskii and Vabischevich 2007). Inequality (1.4) yields to a continuous dependence of the problem's solution on the initial conditions, and hence to well-posedness of the problem in the sense of Tikhonov.

Example 1.4 Consider the problem for the 2-D Laplace equation (similar to the problem presented in Example 1.2):

$$\partial^2 u \left(y, x \right) / \partial y^2 = -\partial^2 u \left(y, x \right) / \partial x^2, \; \left(x, y \right) \in \left(0, 1 \right) \times \left(0, 1 \right),$$

with the conditions

$$u\,(0,x) = f(x), \quad \partial u\,(y,x)\,/\partial y|_{y=0} = 0, \quad x \in [0,1]\,,$$

$$u\,(y,0) = 0 = u\,(y,1)\,, \quad y \in [0,1]\,.$$

The solution of the problem satisfies the inequality

$$\int\limits_0^1 u^2\,(x,y)\,dx \le \left(\int\limits_0^1 f^2(x)dx\right)^{1-y} \left(\int\limits_0^1 u^2\,(x,1)\,dx\right)^y$$

for $|u\,(x,y)| \le M = const$, $(x,y) \in (0,1) \times (0,1)$ (e.g., Kabanikhin 2011). This inequality shows a continuous dependence of the problem on the prescribed conditions, and the problem itself is well-posed in the sense of Tikhonov. Therefore, the Tikhonov's approach allows for developing methods for regularization of the numerical solution of unstable problems.

1.2 Forward and Backward Modelling and Source of Errors

Forward modelling in geodynamics is associated with the solution of direct mathematical problems, and backward modelling with the solution of inverse (time-reverse or retrospective) problems. In forward modelling one starts with unknown initial conditions, which are added to a set of governing equations, a rheological law, and boundary conditions to define properly the relevant mathematical problem. Once the problem is stated, a numerical model (a set of discrete equations) is solved forward in time to obtain future states starting from an initial condition. The initial condition varies (keeping all other model parameters unchanged) to fit model results to reality (observations) as much as possible. Because the model depends on the initial conditions and they are unknown *a priori*, the task "to fit model results to reality" becomes difficult.

Another approach is to use backward modelling. In this case observations are employed as the input condition for a mathematical model. The term "input condition" is used in backward modelling to distinguish it from the initial condition used in forward modelling, although the input condition is an initial condition for the mathematical model in backward modelling. The aim of backward modelling in geodynamics is to find the initial condition in the geological past from available observations.

A numerical model has three kinds of variables: state variables, input variables, and parameters. State variables describe the physical properties of the medium (velocity, pressure, temperature) and depend on time and space. Input variables have to be provided to the model (initial or boundary conditions), most of the time these variables are not directly measured but they can be estimated through data assimilation. Most models contain also a set of parameters (e.g. viscosity, thermal diffusivity), which have to be tuned to adjust the model to the observations. All the variables can be polluted by errors (Ismail-Zadeh and Tackley 2010).

There are three kinds of systematic errors in numerical modelling of geodynamical problems: model, discretisation, and iteration errors. Model errors are associated with the idealization of Earth dynamics by a set of conservation equations governing the dynamics. The model errors are defined as the difference between the actual Earth dynamics and the exact solution of the mathematical model. Discretisation errors are defined as the difference between the exact solution of the conservation equations and the exact solution of the algebraic system of equations obtained by discretising these equations. And iteration errors are defined as the difference between the iterative and exact solutions of the algebraic system of equations. It is important to be aware of the existence of these errors, and even more to try to distinguish one from another.

Apart from the errors associated with the numerical modelling, another two components of errors are essential during data assimilation: (i) data misfit associated with the uncertainties in the distribution of measured physical parameters (e.g. temperature) and (ii) errors associated with the uncertainties in initial and boundary conditions. For example, since there are no direct measurements of temperatures in the Earth's interior, the temperatures can be estimated indirectly from either seismic wave or their anomalies, geochemical analysis or through the extrapolation of surface heat flow observations. Many models of temperature in the lower crust and in the mantle are based on the conversion of seismic tomography data into temperature. Meanwhile, a seismic tomography image of the Earth's mantle is a model indeed and incorporates its own model errors. Another source of uncertainty comes from the choice of geochemical compositions in temperature modelling from the seismic velocities. If data (e.g., synthetic temperature models in the mantle) are biased, information on temperature can be improperly propagated to the past. The temperature at the lower boundary of the model domain used in forward and backward numerical modelling is, of course, an approximation to the real temperature, which is unknown and may change over time at this boundary. Incomplete knowledge of thermal characteristics of the crust and the mantle is another source of errors, which can be propagated into the past during data-driven numerical modelling in geodynamics.

1.3   Data Assimilation Methods

To solve a geodynamic problem, when there are measured/observed data on some physical parameters of the problem but initial and/or some boundary conditions are unknown, data assimilation techniques can be used to constrain the conditions from the data. The initial (boundary) conditions so obtained can then be used to restore geodynamical characteristics of the problem (e.g. temperature, velocity). Data assimilation can be defined as the incorporation of observations and initial/boundary conditions in an explicit dynamic model to provide time continuity and coupling among the physical characteristics of a geodynamic problem.

There are several principal mathematical methods used in quantitative data-driven modelling of geodynamic problems. For example, if heat diffusion can be neglected in a particular problem (e.g. problems of gravity-driven salt tectonics in sedimentary basins), the measured physical parameters can be assimilated into the geological past using the backward advection (BAD) method. Numerical approaches to the solution of the inverse problem of the Rayleigh-Taylor (gravitational) instability were developed for a dynamic restoration of diapiric structures to their earlier stages (e.g., Ismail-Zadeh et al. 2001; Korotkii et al. 2002; Ismail-Zadeh et al. 2004b). Forte and Mitrovica (1997), Steinberger and O'Connell (1997, 1998), Conrad and Gurnis (2003), and Moucha and Forte (2011) modelled the mantle flow backwards in time from present-day mantle density heterogeneities inferred ignoring heat diffusion.

In sequential filtering a numerical model is computed forward in time for the interval for which observations have been made, updating the model each time where observations are available. For example, the sequential filtering was used to compute mantle circulation models (Bunge et al. 2002). Despite sequential data assimilation well adapted to mantle circulation studies, each individual observation influences the model state at later times. Information propagates from the geological past into the future, although our knowledge of the Earth's mantle at earlier times is much poor than that at present.

The variational (VAR) data assimilation method has been pioneered by meteorologists and used very successfully to improve operational weather forecasts (e.g. Kalnay 2003). The data assimilation has also been widely used in oceanography (e.g. Bennett 1992) and in hydrological studies (e.g. McLaughlin 2002). The use of VAR data assimilation in models of mantle dynamics (to estimate mantle temperature and flow in the geological past) has been put forward by Bunge et al. (2003) and Ismail-Zadeh et al. (2003a, b). The major difference between the two approaches are that Bunge et al. (2003) applied the VAR method to the coupled Stokes, continuity, and heat equations (generalized inverse), whereas Ismail-Zadeh et al. (2003a) applied the VAR method to the heat equation only. The VAR approach by Ismail-Zadeh et al. (2003a) is computationally less expensive, because it does not involve the Stokes equation into the iterations between the direct and adjoint problems. Moreover, this approach admits the use of temperature-dependent viscosity.

The VAR data assimilation method was employed for numerical restoration of models of present prominent mantle plumes to their past stages (Ismail-Zadeh et al. 2004a; Hier-Majumder et al. 2005). Effects of thermal diffusion and temperature-dependent viscosity on the evolution of mantle plumes was studied by Ismail-Zadeh et al. (2006) to recover the structure of mantle plumes prominent in the past from that of present plumes weakened by thermal diffusion. Liu and Gurnis (2008) simultaneously inverted mantle properties and initial conditions using the VAR data assimilation method and applied the method to reconstruct the evolution of the

Farallon Plate subduction (Liu et al. 2008). Horbach et al. (2014) demonstrate the
practicality of the method for use in a high resolution mantle circulation model
by restoring a representation of present day mantle heterogeneity derived from
the global seismic shear wave study backward in time for the past 40 million
years (Myr). Worthen et al. (2014) used the VAR (adjoint) method to infer mantle
rheological parameters from surface velocity observations and instantaneous mantle
flow models. Ratnaswamy et al. (2015) developed adjoint-based approach to infer
plate boundary strength and rheological parameters in models of mantle flow from
surface velocity observations, although, compared to Worthen et al. (2014), they
formulated the inverse problem in a Bayesian inference framework. Korotkii et al.
(2016) applied the VAR method to determine thermal and dynamic characteristics
within a lava flow from thermal measurements at lava's surface.

The quasi-reversibility (QRV) method was introduced by Lattes and Lions
(1969). The use of the QRV method implies the introduction into the backward
heat equation of the additional term involving the product of a small regularization
parameter and a higher order temperature derivative. The data assimilation in this
case is based on a search of the best fit between the forecast model state and the
observations by minimizing the regularization parameter. The QRV method was
introduced in geodynamic modelling (Ismail-Zadeh et al. 2007) and employed to
assimilate data in models of lithosphere/mantle dynamics beneath the Carpathian
region (Ismail-Zadeh et al. 2008) and beneath the Japanese islands (Ismail-Zadeh
et al. 2013).

In numerical modelling sensitivity analysis assists in understanding the stability
of the model solution to small perturbations in input variables or parameters. For
instance, if we consider mantle temperature in the geological past as a solution to
the backward model, what will be its variation if there is some perturbation on the
inputs of the model (e.g. present temperature data)? For example, the gradient of the
objective functional with respect to input parameters in variational data assimilation
gives (see Chap. 3) the first-order sensitivity coefficients. Hier-Majumder et al.
(2006) performed the first-order sensitivity analysis for two-dimensional problems
of thermo-convective flow in the mantle. The second-order adjoint sensitivity
analysis presents some challenge associated with cumbersome computations of the
product of the Hessian matrix of the objective functional with a vector (Le Dimet et
al. 2002). See Cacuci (2003) and Cacuci et al. (2005) for more detail on sensitivity
and uncertainty analysis.

References

Aster RC, Borchers B, Thurber CH (2005) Parameter estimation and inverse problems, vol 90,
    International geophysics series. Elsevier, San Diego
Bennett AF (1992) Inverse methods in physical oceanography. Cambridge University Press,
    Cambridge

Bunge H-P, Richards MA, Baumgardner JR (2002) Mantle circulation models with sequential data-assimilation: Inferring present-day mantle structure from plate motion histories. Phil Trans R Soc A 360:2545–2567

Bunge H-P, Hagelberg CR, Travis BJ (2003) Mantle circulation models with variational data assimilation: inferring past mantle flow and structure from plate motion histories and seismic tomography. Geophys J Int 152:280–301

Cacuci DG (2003) Sensitivity and uncertainty analysis. Volume I: theory. Chapman & Hall/CRC, Boca Raton

Cacuci DG, Ionescu-Bujor M, Navon IM (2005) Sensitivity and uncertainty analysis. Volume II: applications to large-scale systems. Chapman & Hall/CRC, Boca Raton

Conrad CP, Gurnis M (2003) Seismic tomography, surface uplift, and the breakup of Gondwanaland: integrating mantle convection backwards in time. Geochem Geophys Geosys 4(3). doi:10.1029/2001GC000299

Forte AM, Mitrovica JX (1997) A resonance in the Earth's obliquity and precession over the past 20 Myr driven by mantle convection. Nature 390:676–680

Hadamard J (1902) Sur les problèmes aux dérivées partielles et leur signification physique. Princeton Univ Bull 13:49–52

Hier-Majumder CA, Belanger E, DeRosier S, Yuen DA, Vincent AP (2005) Data assimilation for plume models. Nonlinear Process Geophys 12:257–267

Hier-Majumder CA, Travis BJ, Belanger E, Richard G, Vincent AP, Yuen DA (2006) Efficient sensitivity analysis for flow and transport in the Earth's crust and mantle. Geophys J Int 166:907–922

Horbach A, Bunge H-P, Oeser J (2014) The adjoint method in geodynamics: derivation from a general operator formulation and application to the initial condition problem in a high resolution mantle circulation model. Int J Geomath 5:163–194

Ismail-Zadeh A, Tackley P (2010) Computational methods for geodynamics. Cambridge University Press, Cambridge

Ismail-Zadeh AT, Talbot CJ, Volozh YA (2001) Dynamic restoration of profiles across diapiric salt structures: numerical approach and its applications. Tectonophysics 337:21–36

Ismail-Zadeh AT, Korotkii AI, Tsepelev IA (2003a) Numerical approach to solving problems of slow viscous flow backwards in time. In: Bathe KJ (ed) Computational fluid and solid mechanics. Elsevier Science, Amsterdam, pp 938–941

Ismail-Zadeh AT, Korotkii AI, Naimark BM, Tsepelev IA (2003b) Three-dimensional numerical simulation of the inverse problem of thermal convection. Comput Math Math Phys 43(4):587–599

Ismail-Zadeh A, Schubert G, Tsepelev I, Korotkii A (2004a) Inverse problem of thermal convection: numerical approach and application to mantle plume restoration. Phys Earth Planet Inter 145:99–114

Ismail-Zadeh AT, Tsepelev IA, Talbot CJ, Korotkii AI (2004b) Three-dimensional forward and backward modelling of diapirism: numerical approach and its applicability to the evolution of salt structures in the Pricaspian basin. Tectonophysics 387:81–103

Ismail-Zadeh A, Schubert G, Tsepelev I, Korotkii A (2006) Three-dimensional forward and backward numerical modeling of mantle plume evolution: effects of thermal diffusion. J Geophys Res 111:B06401. doi:10.1029/2005JB003782

Ismail-Zadeh A, Korotkii A, Schubert G, Tsepelev I (2007) Quasi-reversibility method for data assimilation in models of mantle dynamics. Geophys J Int 170:1381–1398

Ismail-Zadeh A, Schubert G, Tsepelev I, Korotkii A (2008) Thermal evolution and geometry of the descending lithosphere beneath the SE-Carpathians: an insight from the past. Earth Planet Sci Lett 273:68–79

Ismail-Zadeh A, Honda S, Tsepelev I (2013) Linking mantle upwelling with the lithosphere descent and the Japan Sea evolution: a hypothesis. Sci Rep 3:1137

Kabanikhin SI (2011) Inverse and ill-posed problems. Theory and applications. De Gruyter, Berlin

Kalnay E (2003) Atmospheric modeling, data assimilation and predictability. Cambridge University Press, Cambridge

Kirsch A (1996) An introduction to the mathematical theory of inverse problems. Springer, New York

Korotkii AI, Tsepelev IA, Ismail-Zadeh AT, Naimark BM (2002) Three-dimensional backward modeling in problems of Rayleigh-Taylor instability. Proc Ural State Univ 22:96–104 (in Russian)

Korotkii A, Kovtunov D, Ismail-Zadeh A, Tsepelev I, Melnik O (2016) Quantitative reconstruction of thermal and dynamic characteristics of lava flow from surface thermal measurements. Geophys J Int. doi:10.1093/gji/ggw117

Lattes R, Lions JL (1969) The method of quasi-reversibility: applications to partial differential equations. Elsevier, New York

Le Dimet F-X, Navon IM, Daescu DN (2002) Second-order information in data assimilation. Mon Weather Rev 130:629–648

Liu L, Gurnis M (2008) Simultaneous inversion of mantle properties and initial conditions using an adjoint of mantle convection. J Geophys Res 113:B08405. doi:10.1029/2008JB005594

Liu L, Spasojevic S, Gurnis M (2008) Reconstructing Farallon plate subduction beneath North America back to the Late Cretaceous. Science 322:934–938

McLaughlin D (2002) An integrated approach to hydrologic data assimilation: interpolation, smoothing, and forecasting. Adv Water Res 25:1275–1286

Moucha R, Forte AM (2011) Changes in African topography driven by mantle convection. Nat Geosci 4:707–712

Ratnaswamy V, Stadler G, Gurnis M (2015) Adjoint-based estimation of plate coupling in a nonlinear mantle flow model: theory and examples. Geophys J Int 202:768–786

Samarskii AA, Vabishchevich PN (2007) Numerical methods for solving inverse problems of mathematical physics. De Gruyter, Berlin

Steinberger B, O'Connell RJ (1997) Changes of the Earth's rotation axis owing to advection of mantle density heterogeneities. Nature 387:169–173

Steinberger B, O'Connell RJ (1998) Advection of plumes in mantle flow: implications for hotspot motion, mantle viscosity and plume distribution. Geophys J Int 132:412–434

Tikhonov AN (1963) Solution of incorrectly formulated problems and the regularization method. Dokl Akad Nauk SSSR 151:501–504 (Engl. transl.: Soviet Math Dokl 4:1035–1038)

Turcotte DL, Schubert G (2002) Geodynamics, 2nd edn. Cambridge University Press, Cambridge

Worthen J, Stadler G, Petra N, Gurnis M, Ghattas O (2014) Towards an adjoint-based inversion for rheological parameters in nonlinear viscous mantle flow. Phys Earth Planet Int 234:23–34

Chapter 2
Backward Advection Method and Its Application to Modelling of Salt Tectonics

Abstract This chapter deals with the simplest method in data assimilation allowing for solving a geodynamic problem backward in time by suppressing thermal diffusion. The method is suitable in the advection-dominated regimes of thermal convective flows. We present an application of the method to three-dimensional numerical modelling of salt diapirism in sedimentary basins.

Keywords Backward advection • Salt buoyancy • Diapir • Sedimentary basin • Restoration • Numerical modelling

2.1 Basic Idea of the Backward Advection (BAD) Method

The principal mathematical difficulty in solving some geodynamic problems (e.g. thermal convective flows in the mantle) backward in time is the ill-posedness of the backward heat problem and the presence of the heat diffusion term in the heat equation (Kirsch 1996). The backward advection (BAD) method suggests neglecting the thermal diffusion term, and the heat advection equation can then be solved backward in time. In the case of advection-dominated fluid flows with an insignificant diffusion, this approach is valid.

Both direct (forward in time) and inverse (backward in time) advection problems are well-posed. This is because the time-dependent advection equation has the same form of characteristics for the direct and inverse velocity field: the vector velocity reverses its direction, when time is reversed. Therefore, numerical algorithms used to solve the direct problem can also be used in studies of the time-reverse problems by replacing positive time steps with negative ones.

Using the BAD method, Steinberger and O'Connell (1998) studied the motion of hotspots relative to the deep mantle. They combined the advection of plumes, which are thought to cause the hotspots on the Earth's surface, with a large-scale mantle flow field and constrained the viscosity structure of the Earth's mantle. Conrad and Gurnis (2003) modelled the history of mantle flow using a tomographic image of the mantle beneath southern Africa as an input (initial) condition for the backward mantle advection model while reversing the direction of flow. If the resulting model of the evolution of thermal structures obtained by the BAD method is used as a starting point for a forward mantle convection model, present mantle structures can be reconstructed if the time of assimilation does not exceed 50–75 Myr. Moucha and Forte (2011) simulated mantle convection using the BAD method to reconstruct the evolution of dynamic topography of Africa over the past 30 Myr.

In what follows we present the application of the BAD method to solving inverse problem of the gravitational, or the Rayleigh-Taylor (RT), instability, namely, the dynamic restoration of diapiric salt structures to their earlier stages of the evolution studied by Ismail-Zadeh et al. (2004b).

2.2   Modelling of Salt Diapirism

Salt is so buoyant and weak compared to most other rocks with which it is found that it develops distinctive structures with a wide variety of shapes and relationships with other rocks by various combinations of gravity, thermal effects, and lateral forces. The crests of passive salt bodies can stay near the sedimentation surface while their surroundings are buried (downbuilt) by other sedimentary rocks (Jackson et al. 1994). The profiles of downbuilt passive diapirs can simulate those of fir trees because they reflect the ratio of increase in diapir height relative to the rate of accumulation of the downbuilding sediments (Talbot 1995) and lateral forces (Koyi 1996). Salt movements can be triggered by faulting and driven by erosion and redeposition, differential loading, buoyancy and other geological processes. For example, differential loading and buoyancy are proposed as primary driving mechanisms for salt tectonics in the Pricaspian Basin (Volozh et al. 2003), whereas faulting, erosion, and buoyancy are considered as principal mechanisms responsible for salt movements in the Dnieper-Donets Basin (Lobkovsky et al. 1996; Stovba and Stephenson 2002). Many salt sequences are buried by overburdens sufficiently stiff to resist the buoyancy of the salt. Such salt will only be driven by differential loading into sharp-crested reactive-diapiric walls after the stiff overburden is weakened and thinned by faults. Such reactive diapirs often rise up and out of the fault zone and thereafter can continue increasing in relief as by passive downbuilding of more sediment.

Active diapirs are those that lift or displace their overburdens. Although any erosion of the crests of salt structures and deposition of surrounding overburden rocks influence their growth, diapirs with significant relief have sufficient buoyancy to rise (upbuild) through stiff overburdens (Jackson et al. 1994). The rapid deposition of denser and more viscous sediments over less dense and viscous salt results in the RT instability. This leads to a gravity-driven single overturn of the salt layer with its denser but ductile overburden. RT overturns (Ramberg 1968) are characterized by the rise of rocksalt through overlying and younger compacting clastic sediments that are deformed as a result. The consequent salt structures evolve through a great variety of shapes. Perturbations of the interface between salt and its denser overburden result in the overburden subsiding as salt rises owing to the density inversion (Ismail-Zadeh et al. 2002).

Two-dimensional (2-D) numerical models of salt diapirism examined how the viscosity ratio between the salt and its overburden affects the shapes and growth rate of diapirs (Woidt 1978). Schmeling (1987) demonstrated how the dominant wavelength and the geometry of gravity overturns are influenced by the initial shape of the interface between the salt and its overburden. Later Poliakov et al. (1993a) and Naimark et al. (1998) developed numerical models of diapiric growth considering the effects of sedimentation and redistribution of sediments. Van Keken et al. (1993), Poliakov et al. (1993b), and Daudre and Cloetingh (1994) introduced non-linear rheological properties of salt and overburden into their numerical models.

Two-dimensional analyses of the evolution of salt structures are restricted and not suitable for examining the complicated shapes of mature diapiric patterns. Resolving the geometry of gravity overturns requires three-dimensional (3-D) numerical modelling. Ismail-Zadeh et al. (2004a) analysed such typical 3-D structures as deep polygonal buoyant ridges, shallow salt-stock canopies, and salt walls. The increasing application of 3-D seismic exploration in oil and gas prospecting points to the need for vigorous efforts toward numerical modelling of the evolution of salt structures in three dimensions, both forwards and backwards in time.

Most numerical models of salt diapirism involved the forward evolution of salt structures toward increasing maturity. Ismail-Zadeh et al. (2001a) developed a numerical approach to 2-D dynamic restoration of cross-sections across salt structures. The approach was based on solving the inverse problem of RT instability by the BAD method and simultaneous back-stripping of uppermost sediments. The same method was used in 3-D cases to model RT instability backward in time (Korotkii et al. 2002; Ismail-Zadeh et al. 2004b).

2.3   Mathematical Statement

The advection problem (a gravity flow of an incompressible fluid of variable density and viscosity) is considered in the rectangular domain $\Omega = [0, x_1 = 3h] \times [0, x_2 = 3h] \times [0, x_3 = h]$, where $\mathbf{x} = (x_1, x_2, x_3)$ are the Cartesian coordinates and h is the depth of the domain. The following governing equations describe the slow movement of salt and its overburden (e.g., Ismail-Zadeh et al. 2004b):

momentum conservation

$$\nabla P = \nabla \cdot \left[\eta \left(\nabla \mathbf{u} + \nabla \mathbf{u}^T \right) \right] + \mathbf{F} \tag{2.1}$$

continuity for incompressible fluid

$$\nabla \cdot \mathbf{u} = \partial u_1 / \partial x_1 + \partial u_2 / \partial x_2 + \partial u_3 / \partial x_3 = 0 \tag{2.2}$$

and advection of density and viscosity with the flow

$$\partial \rho / \partial t + \langle \mathbf{u}, \nabla \rho \rangle = 0, \quad \partial \eta / \partial t + \langle \mathbf{u}, \nabla \eta \rangle = 0 \tag{2.3}$$

Equations (2.1), (2.2) and (2.3) contain the following variables and parameters: time t; velocity $\mathbf{u} = (u_1(t, \mathbf{x}), u_2(t, \mathbf{x}), u_3(t, \mathbf{x}))$; pressure $P = P(t, \mathbf{x})$; density $\rho = \rho(t, \mathbf{x})$; viscosity $\eta = \eta(t, \mathbf{x})$; and the body force per unit volume $\mathbf{F} = (0, 0, -g\rho)$, where g is the acceleration due to gravity. Here, ∇ and $\nabla \cdot$ denote the gradient and divergence operators, respectively; $\mathbf{E} \equiv \nabla \mathbf{u} + \nabla \mathbf{u}^T$ is the strain rate tensor $\mathbf{E} = \{e_{ij}(\mathbf{u})\} = \{\partial u_i/\partial x_j + \partial u_j/\partial x_i\}$, $\nabla \cdot (\eta \mathbf{E}) = \left(\sum_{m=1}^{3} \frac{\partial(\eta e_{m1})}{\partial x_m}, \sum_{m=1}^{3} \frac{\partial(\eta e_{m2})}{\partial x_m}, \sum_{m=1}^{3} \frac{\partial(\eta e_{m3})}{\partial x_m} \right)$; and $\langle \cdot, \cdot \rangle$ denotes the scalar product of vectors. Equations (2.1), (2.2) and (2.3) make up a set of equations that determine the unknown $\mathbf{u}$, P, ρ, and η as functions of independent variables t and $\mathbf{x}$.

The number of unknowns is reduced by introducing the two-component representation of the velocity potential $\boldsymbol{\Psi} = (\psi_1, \psi_2, \psi_3 = 0)$ (Ismail-Zadeh et al. 2001b), from which the velocity is obtained as

$$\mathbf{u} = \text{curl } \boldsymbol{\Psi}; \quad u_1 = -\frac{\partial \psi_2}{\partial x_3}, \quad u_2 = -\frac{\partial \psi_1}{\partial x_3}, \quad u_3 = \frac{\partial \psi_2}{\partial x_1} - \frac{\partial \psi_1}{\partial x_2}. \tag{2.4}$$

Applying the curl operator to (2.1) and using the identities $\text{curl} (\nabla P) \equiv 0$ and $\nabla \cdot (\text{curl } \boldsymbol{\Psi}) \equiv 0$, the following equations can be derived from (2.1) and (2.2):

$$\begin{aligned} D_{2i}(\eta e_{i3}) - D_{3i}(\eta e_{i2}) &= gD_2\rho, \\ D_{3i}(\eta e_{i1}) - D_{1i}(\eta e_{i3}) &= -gD_1\rho, \\ D_{1i}(\eta e_{i2}) - D_{2i}(\eta e_{i1}) &= 0, \quad i = 1, 2, 3 \end{aligned} \tag{2.5}$$

where $D_{ij} = \partial^2/\partial x_i \partial x_j$, $D_i = \partial/\partial x_i$, and a summation over repeated subscripts is assumed hereinafter. The strain rate components e_{ij} are defined in terms of the vector velocity potential as

$$\begin{aligned} e_{11} &= -2D_{13}\psi_2, \quad e_{22} = 2D_{23}\psi_1, \quad e_{33} = 2(D_{31}\psi_2 - D_{32}\psi_1), \\ e_{12} &= e_{21} = D_{13}\psi_1 - D_{23}\psi_2, \quad e_{13} = e_{31} = D_{11}\psi_2 - D_{33}\psi_2 - D_{12}\psi_1, \\ e_{23} &= e_{32} = D_{33}\psi_1 - D_{22}\psi_1 + D_{21}\psi_2. \end{aligned} \tag{2.6}$$

At the initial time $t_0 = 0$ the density and viscosity are assumed to be known. On the boundary Γ of Ω, which consists of the faces $x_i = 0$ and $x_i = l_i$ ($i = 1, 2, 3$), the condition of impenetrability with perfect slip is imposed:

$$\langle \mathbf{u}, \mathbf{n} \rangle = 0, \quad \sigma \mathbf{n} - \langle \sigma \mathbf{n}, \mathbf{n} \rangle \mathbf{n} = 0, \tag{2.7}$$

where $\mathbf{n}$ is the outward unit normal vector at a point on the boundary, and $\sigma = \{\sigma_{ij}\} = \eta \{e_{ij}\}$ is the stress tensor.

In terms of the vector velocity potential the boundary conditions (2.7) take the following forms:

$$\psi_2 = D_1\psi_1 = D_{11}\psi_2 = 0 \qquad \text{at } \Gamma_1 (x_1 = 0) \text{ and } \Gamma_1 (x_1 = l_1),$$
$$\psi_1 = D_2\psi_2 = D_{22}\psi_1 = 0 \qquad \text{at } \Gamma_2 (x_2 = 0) \text{ and } \Gamma_2 (x_2 = l_2),$$
$$\psi_1 = \psi_2 = D_{33}\psi_1 = 0 \qquad \text{at } \Gamma_3 (x_3 = 0) \text{ and } \Gamma_3 (x_3 = l_3). \tag{2.8}$$

Thus, the problem of gravitational advection is to determine functions $\psi_1 = \psi_1 (t, \mathbf{x})$, $\psi_2 = \psi_2 (t, \mathbf{x})$, $\rho = \rho (t, \mathbf{x})$, and $\eta = \eta (t, \mathbf{x})$ satisfying (2.3) and (2.5) in Ω at $t \geq t_0$, the prescribed boundary (2.8) and the initial conditions.

2.4   Solution Method

To solve a set of equations (2.5) numerically, an Eulerian FEM (Ismail-Zadeh and Tackley 2010) is employed, and these equations are replaced by an equivalent variational equation. Namely, consider any arbitrary admissible test vector function $\mathbf{\Phi} = (\varphi_1, \varphi_2, \varphi_3 = 0)$ satisfying the same conditions as for the vector function $\mathbf{\Psi}$ and multiply the first two equations of Eq. (2.5) by φ_1 and φ_2, respectively. Taking the result and integrating by parts over Ω, and using the boundary conditions for the desired and test vector functions, the following variational equation is obtained

$$\aleph (\eta; \mathbf{\Psi}, \mathbf{\Phi}) = \Re (\eta, \rho; \mathbf{\Phi}), \tag{2.9}$$

where

$$\aleph (\eta; \mathbf{\Psi}, \mathbf{\Phi}) = \iiint_\Omega \eta \left[2e_{11}\tilde{e}_{11} + 2e_{22}\tilde{e}_{22} + 2e_{33}\tilde{e}_{33} + e_{12}\tilde{e}_{12} + e_{13}\tilde{e}_{13} + e_{23}\tilde{e}_{23} \right] d\mathbf{x},$$

$$\Re (\eta, \rho; \mathbf{\Phi}) = \iiint_\Omega g\rho \left(\frac{\partial \varphi_1}{\partial x_2} - \frac{\partial \varphi_2}{\partial x_1} \right) d\mathbf{x},$$

and the expressions for $\tilde{e}_{ij}$ in terms of $\mathbf{\Phi}$ are identical to the expressions for e_{ij} in terms of the function $\mathbf{\Psi}$.

The components of the vector velocity potential are represented as a sum of tri-cubic splines

$$\psi_s (t, \mathbf{x}) \approx \psi_{ijk}^s(t)\gamma_i^s (x_1) \zeta_j^s (x_2) \vartheta_k^s (x_3), \quad s = 1, 2 \tag{2.10}$$

with the unknown functions $\psi_{ijk}^s(t)$ (see Fig. 4.6 in Ismail-Zadeh and Tackley 2010 for the representation of the basic splines). Hereinafter, $i, l, p = 1, 2, \ldots, N_1$; $j, m, q = 1, 2, \ldots, N_2$; and $k, n, r = 1, 2, \ldots, N_3$. Density and viscosity are approximated by linear combinations of appropriate tri-linear basis functions:

$$\rho (t, \mathbf{x}) \approx \rho_{ijk}(t)\tilde{s}_i^1 (x_1) \tilde{s}_j^2 (x_2) \tilde{s}_k^3 (x_3), \quad \eta (t, \mathbf{x}) \approx \eta_{ijk}(t)\tilde{s}_i^1 (x_1) \tilde{s}_j^2 (x_2) \tilde{s}_k^3 (x_3), \tag{2.11}$$

where $\tilde{s}_i^1\,(x_1)$, $\tilde{s}_j^2\,(x_2)$, and $\tilde{s}_k^3\,(x_3)$ are linear basis functions. The trilinear basis functions provide good approximations for step functions (such as density or viscosity that change abruptly from one layer to another).

Substituting approximations (2.9), (2.10) into the variational Eq. (2.8), the following system of linear algebraic equations (SLAE) for the unknown $\psi_{ijk}^s(t)$ is obtained:

$$
C_{sijk}^{lmn}\left(\eta_{ijk}\right)\begin{pmatrix}\psi_{ijk}^1\\\psi_{ijk}^2\end{pmatrix}=g\rho_{ijk}\begin{pmatrix}P_{il}^{01}&Q_{jm}^{00}&R_{kn}^{00}\\-P_{il}^{00}&Q_{jm}^{01}&R_{kn}^{00}\end{pmatrix}. \tag{2.12}
$$

The coefficients $C_{sijk}^{lmn}=\displaystyle\sum_{a_1a_2b_1b_2c_1c_2}\sum_{p,q,r}\eta_{pqr}w_{a_1a_2b_1b_2c_1c_2}A_{silp}^{a_1a_2}B_{sjmq}^{b_1b_2}C_{sknr}^{c_1c_2}$ in (2.11) are the integrals of various products of basic functions (the cubic splines) and their derivatives. Here $a_1+b_1+c_1=2$, $a_2+b_2+c_2=2$; the values of $w_{a_1a_2b_1b_2c_1c_2}$ are readily obtained by collecting similar terms in the sums; and coefficients $A_{silp}^{a_1a_2}$, $B_{sjmq}^{b_1b_2}$, and $C_{sknr}^{c_1c_2}$ are integrals of the form

$$
A_{silp}^{a_1a_2}=\int_0^{l_1}\left(D_{a_1}\gamma_i^s\,(x_1)\right)\left(D_{a_2}\gamma_l^s\,(x_1)\right)\tilde{s}_p^1\,(x_1)\,dx_1,
$$

$$
B_{sjmq}^{b_1b_2}=\int_0^{l_2}\left(D_{b_1}\zeta_j^s\,(x_2)\right)\left(D_{b_2}\zeta_m^s\,(x_2)\right)\tilde{s}_q^2\,(x_2)\,dx_2,
$$

$$
C_{sknr}^{c_1c_2}=\int_0^{l_3}\left(D_{c_1}\vartheta_k^s\,(x_3)\right)\left(D_{c_2}\vartheta_n^s\,(x_3)\right)\tilde{s}_r^3\,(x_3)\,dx_3,
$$

where $\{\gamma\}$, $\{\zeta\}$, and $\{\vartheta\}$ are the cubic basis splines and $\{\tilde{s}\}$ are linear basis functions. Coefficients in the right-hand side of (2.11) are represented as:

$$
P_{il}^{ab}=\int_0^{l_1}\left(D_a\tilde{s}_i^1\,(x_1)\right)\left(D_b\gamma_l^1\,(x_1)\right)dx_1,
$$

$$
Q_{jm}^{ab}=\int_0^{l_2}\left(D_a\tilde{s}_j^2\,(x_2)\right)\left(D_b\zeta_m^1\,(x_2)\right)dx_2,
$$

$$
R_{kn}^{ab}=\int_0^{l_3}\left(D_a\tilde{s}_k^3\,(x_3)\right)\left(D_b\vartheta_n^1\,(x_3)\right)dx_3.
$$

The SLAE is solved by the conjugate gradient method designed for multi-processor computers; approximations of the density and viscosity for a prescribed velocity can be computed by the method of characteristics (see, e.g., Ismail-Zadeh and Tackley 2010). The accuracy of the numerical method was tested by Ismail-Zadeh et al. (2001b) using the analytical solution to the coupled Stokes and density advection equations (Trushkov 2002), and verifying the conservation of mass at each time step, and the accuracy of the vector velocity potential $\boldsymbol{\Psi}$.

2.5 Forward and Backward Model Results

Although dimensionless values and functions are used in computations, numerical results are presented below in dimensional form for the reader's convenience. The time step Δt is chosen from the condition that the maximum displacement does not exceed a given small value h: $\Delta t = h/u_{max}$, where u_{max} is the maximum value of the flow velocity. The model domain is a rectangular region ($h = 10$ km) divided into $38 \times 38 \times 38$ rectangular elements in order to approximate the vector velocity potential and viscosity. Density is represented on a grid three times finer, $112 \times 112 \times 112$. The model viscosities and densities are assumed to be 10^{20} Pa s and 2.65×10^3 kg m^{-3} for the overburden layer and 10^{18} Pa s and 2.24×10^3 kg m^{-3} for the salt layer, respectively.

Salt diapirs in the numerical model evolve from random initial perturbations of the interface between the salt and its overburden deposited on the top of horizontal salt layer prior to the interface perturbation. A salt layer of 3 km thick at the bottom of the model box is overlain by a sedimentary overburden of 7 km thick at time $t = 0$. The interface between the salt and its overburden was disturbed randomly with an amplitude ~100 m. Figures 2.1a–d, a front view and 2.2a–d, a top view show the positions of the interface between salt and overburden in the model at successive times over a period of about 21 Myr. The evolution clearly shows two major phases: an initial phase resulting in the development of salt pillows lasting about 18 Myr (a, b) and a mature phase resulting in salt dome evolution lasting about 3 Myr (c, d).

To restore the evolution of salt diapirs predicted by the forward model through successive earlier stages, a positive time is replaced by a negative time, and the problem is solved backward in time. Such a replacement is possible, because the characteristics of the advection equations have the same form for both direct and inverse velocity field. The final position of the interface between salt and its overburden in the forward model (Figs. 2.1d and 2.2d) is used as an initial position of the interfaces for the backward model. Figures 2.1d–g and 2.2d–g illustrate successive steps in the restoration of the upbuilt diapirs. Least square errors δ of the restoration are calculated using the formula:

$$\delta(x_1, x_2) = \left(\int_0^h (\rho(x_1, x_2, x_3) - \tilde{\rho}(x_1, x_2, x_3))^2 dx_3 \right)^{1/2},$$

where $\rho(x_1, x_2, x_3)$ is the density at initial time, and $\tilde{\rho}(x_1, x_2, x_3)$ is the restored density (Fig. 2.2h). The maximum value δ does not exceed 120 kg m^{-3}, and the error is associated with small areas of the initial interface's perturbation.

To demonstrate the stability of the restoration results with respect to changes in the density of the overburden, the restoration procedure was tested by synthetic examples. Initially the forward model is run for 200 computational time steps (about 30 Myr). Then the density contrast ($\delta\rho$) between salt and its overburden is changed by a few per cent: namely, $\delta\rho$ was chosen to be 400, 405, 410 (the actual contrast),

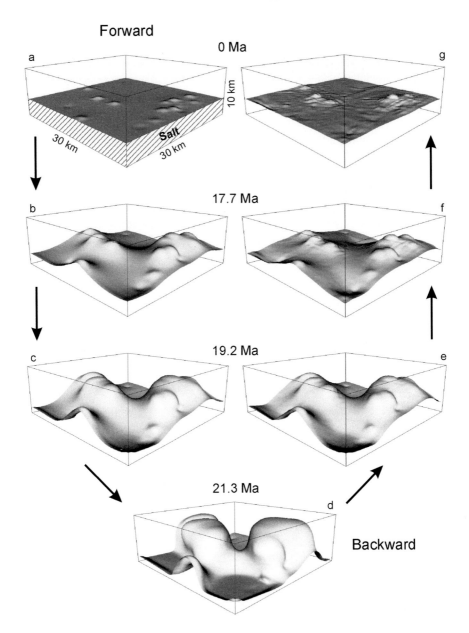

Fig. 2.1 Evolution (*front view*) of salt diapirs toward increasing maturity (**a–d**) and restoration of the evolution (**d–g**). Interfaces between salt and its overburden are presented at successive times. After Ismail-Zadeh et al. (2004b)

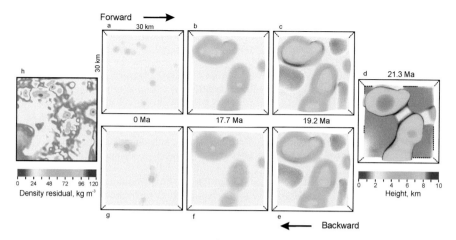

Fig. 2.2 Evolution (*top view*) of salt diapirs toward increasing maturity (**a–d**) and restoration of the evolution (**d–g**) at the same times as in Fig. 2.1. (**h**) Restoration errors (After Ismail-Zadeh et al. 2004b)

415, and 420 kg m^{-3}. The evolution of the system was restored for these density contrasts. Ismail-Zadeh et al. (2004b) found small discrepancies (less than 0.5 %) between least square errors for all these test cases. The tests show that the solution is stable to small changes in the initial conditions, and this is in agreement with the mathematical theory of well-posed problems (Tikhonov and Samarskii 1990). Meanwhile it should be mentioned that if the model is computed for a very long time and the less dense salt layer spreads uniformly into a horizontal layer near the surface, practical restoration of the layered structure becomes impossible (Ismail-Zadeh et al. 2001a).

In this chapter we have discussed the applicability of the BAD method to dynamic restoration of salt structures and their overburden in an isothermal case. Meanwhile temperature affects the maturation of hydrocarbons, and a joint thermal and dynamic restoration would be of significant interest. One of the possibilities is to use dynamic restoration of sedimentary layers and subsequent thermal modelling of the restored layers (e.g., Ismail-Zadeh et al. 2008, 2010) to determinate pressure and temperature conditions in the geological past. Alternatively, more advanced methods of data assimilation, as variational and/or quasi-reversibility methods, can be employed.

References

Conrad CP, Gurnis M (2003) Seismic tomography, surface uplift, and the breakup of Gondwanaland: integrating mantle convection backwards in time. Geochem Geophys Geosys 4. doi:10.1029/2001GC000299

Daudre B, Cloetingh S (1994) Numerical modelling of salt diapirism: influence of the tectonic regime. Tectonophysics 240:59–79

Ismail-Zadeh A, Tackley P (2010) Computational methods for geodynamics. Cambridge University Press, Cambridge

Ismail-Zadeh AT, Talbot CJ, Volozh YA (2001a) Dynamic restoration of profiles across diapiric salt structures: numerical approach and its applications. Tectonophysics 337:21–36

Ismail-Zadeh AT, Korotkii AI, Naimark BM, Tsepelev IA (2001b) Numerical modelling of three-dimensional viscous flow with gravitational and thermal effects. Comput Math Math Phys 41:1331–1345

Ismail-Zadeh AT, Huppert HE, Lister JR (2002) Gravitational and buckling instabilities of a rheologically layered structure: implications for salt diapirism. Geophys J Int 148:288–302

Ismail-Zadeh AT, Tsepelev IA, Talbot C, Oster P (2004a) Three-dimensional modeling of salt diapirism: a numerical approach and algorithm of parallel calculations. Comput Seismol Geodyn 6:33–41

Ismail-Zadeh AT, Tsepelev IA, Talbot CJ, Korotkii AI (2004b) Three-dimensional forward and backward modelling of diapirism: numerical approach and its applicability to the evolution of salt structures in the Pricaspian basin. Tectonophysics 387:81–103

Ismail-Zadeh A, Wilhelm H, Volozh Y (2008) Geothermal evolution of the Astrakhan arch region of the Pricaspian Basin. Int J Earth Sci 97:1029–1043

Ismail-Zadeh A, Wilhelm H, Volozh Y, Tinakin O (2010) The Astrakhan Arch of the Pricaspian Basin: geothermal analysis and modelling. Basin Res 22:751–764

Jackson MPA, Vendeville BC, Schultz-Ela DD (1994) Structural dynamics of salt systems. Annu Rev Earth Planet Sci 22:93–117

Kirsch A (1996) An introduction to the mathematical theory of inverse problems. Springer, New York

Korotkii AI, Tsepelev IA, Ismail-Zadeh AT, Naimark BM (2002) Three-dimensional backward modeling in problems of Rayleigh-Taylor instability. Proc Ural State Univ 22(4):96–104 (in Russian)

Koyi H (1996) Salt flow by aggrading and prograding overburdens. In: Alsop I, Blundell D, Davison I (eds) Salt tectonics. Geological Society Special Publication 100, London, pp 243–258

Lobkovsky LI, Ismail-Zadeh AT, Krasovsky SS, Kuprienko PY, Cloetingh S (1996) Origin of gravity anomalies and possible forming mechanism of the Dnieper-Donets Basin. Tectonophysics 268:281–292

Moucha R, Forte AM (2011) Changes in African topography driven by mantle convection. Nat Geosci 4:707–712

Naimark BM, Ismail-Zadeh AT, Jacoby WR (1998) Numerical approach to problems of gravitational instability of geostructures with advected material boundaries. Geophys J Int 134:473–483

Poliakov ANB, van Balen R, Podladchikov Y, Daudre B, Cloetingh S, Talbot C (1993a) Numerical analysis of how sedimentation and redistribution of surficial sediments affects salt diapirism. Tectonophysics 226:199–216

Poliakov ANB, Podladchikov Y, Talbot C (1993b) Initiation of salt diapirs with frictional overburdens: numerical experiments. Tectonophysics 228:199–210

Ramberg H (1968) Instability of layered system in the field of gravity. Phys Earth Planet Inter 1:427–474

Schmeling H (1987) On the relation between initial conditions and late stages of Rayleigh-Taylor instabilities. Tectonophysics 133:65–80

Steinberger B, O'Connell RJ (1998) Advection of plumes in mantle flow: implications for hotspot motion, mantle viscosity and plume distribution. Geophys J Int 132:412–434

Stovba SM, Stephenson RA (2002) Style and timing of salt tectonics in the Dniepr-Donets Basin (Ukraine): implications for triggering and driving mechanisms of salt movement in sedimentary basins. Mar Petrol Geol 19:1169–1189

Talbot CJ (1995) Molding of salt diapirs by stiff overburdens. In: Jackson MPA, Roberts DG, Snelson S (eds) Salt tectonics – a global perspective. American Association of Petroleum Geologists, Memoir 65, Tulsa, pp 61–75

Tikhonov AN, Samarskii AA (1990) Equations of mathematical physics. Dover Publications, New York

Trushkov VV (2002) An example of $(3+1)$-dimensional integrable system. Acta Appl Math 62:111–122

Van Keken PE, Spiers CJ, van den Berg AP, Muyzert EJ (1993) The effective viscosity of rocksalt: implementation of steady-state creep laws in numerical models of salt diapirism. Tectonophysics 225:457–476

Volozh YA, Talbot CJ, Ismail-Zadeh AT (2003) Salt structures and hydrocarbons in the Pricaspian Basin. Am Assoc Pet Geol Bull 87:313–334

Woidt W-D (1978) Finite element calculations applied to salt dome analysis. Tectonophysics 50:369–386

Chapter 3
Variational Method and Its Application to Modelling of Mantle Plume Evolution

Abstract In this chapter, we present a variational (VAR) method for assimilation of data related to models of thermal convective flow. This approach is based on a search for model parameters (e.g., mantle temperature and flow velocity in the past) by minimizing the differences between present-day observations of the relevant physical parameters (e.g., temperature derived from seismic tomography, geodetic measurements) and those predicted by forward models for an initial guess temperature. To demonstrate the applicability of this method, we present a numerical model of the evolution of mantle plumes and show that the initial shape of the plumes can be accurately reconstructed. Finally we discuss some challenges in the VAR data assimilation including a smoothness of data.

Keywords Variational method • Adjoint problem • Numerical modelling • Mantle plume • Data smoothness • Noise

3.1 Basic Idea of the Variational (VAR) Method

The variational data assimilation is based on a search of the best fit between the forecast model state and the observations by minimizing an objective functional (a normalized residual between the target model and observed variables) over space and time. To minimize the objective functional over time, an assimilation time interval is defined and an adjoint model is typically used to find the derivatives of the objective functional with respect to the model states.

The VAR method (sometimes referred to as the adjoint method) can be formulated with a weak constraint (so-called, a generalized inverse), where errors in the model formulation are taken into account (Bunge et al. 2003), or with a strong constraint where the model is assumed to be perfect except for the errors associated with the initial conditions (Ismail-Zadeh et al. 2003). The generalized inverse of mantle convection considers model errors, data misfit and the misfit of parameters as control variables. As the generalized inverse presents a computational challenge, Bunge et al. (2003) considered a simplified generalized inverse imposing a strong constraint on errors (ignoring all errors except for the initial condition errors). Therefore, the strong constraint makes the problem computationally tractable.

3.2 Mathematical Statement

Although the mantle rheology is more complex (e.g., Karato 2010), we assume here that the mantle behaves as a Newtonian incompressible fluid with a temperature-dependent viscosity and infinite Prandtl number (the dimensionless parameter denoting the ratio between the viscosity and the product of the density and the thermal diffusivity). The mantle flow is described by heat, motion, and continuity equations (Chandrasekhar 1961). To simplify the governing equations, the Boussi-nesq approximation (Boussinesq 1903) is used by keeping the density constant everywhere except for buoyancy term in the equation of motion.

In the model domain $\Omega = [0, x_1 = 3h] \times [0, x_2 = 3h] \times [0, x_3 = h]$, where $\mathbf{x} = (x_1, x_2, x_3)$ are the Cartesian coordinates and h is the depth of the domain, the dimensionless equations take the form:

$$\partial T/\partial t + \langle \mathbf{u}, \nabla T \rangle = \nabla^2 T, \ \mathbf{x} \in \Omega, \ t \in (0, \vartheta), \tag{3.1}$$

$$\nabla P = \nabla \cdot \left[\eta \left(\nabla \mathbf{u} + \nabla \mathbf{u}^T \right) \right] + Ra T \mathbf{e}, \ \mathbf{e} = (0, 0, 1), \tag{3.2}$$

$$\nabla \cdot \mathbf{u} = 0, \ t \in (0, \vartheta), \ \mathbf{x} \in \Omega. \tag{3.3}$$

Here T, t, $\mathbf{u} = (u_1, u_2, u_3)$, P, and η are dimensionless temperature, time, velocity, pressure, and viscosity, respectively. The Rayleigh number is defined as $Ra = \alpha g \rho_{ref} \Delta T h^3 \eta_{ref}^{-1} \kappa^{-1}$, where α is the thermal expansivity, g is the acceleration due to gravity, ρ_{ref} and η_{ref} are the reference typical density and viscosity, respectively; ΔT is the temperature contrast between the lower and upper boundaries of the model domain; and κ is the thermal diffusivity. In Eqs. (3.1), (3.2) and (3.3) length, temperature, and time are normalized by h, ΔT, and $h^2 \kappa^{-1}$, respectively.

At the boundary Γ of the model domain Ω the impenetrability condition and no-slip or perfect slip conditions are prescribed: $\mathbf{u} = 0$ or $\langle \mathbf{u}, \mathbf{n} \rangle = 0$, $\sigma \mathbf{n} - \langle \sigma \mathbf{n}, \mathbf{n} \rangle \mathbf{n} = 0$, where $\mathbf{n}$ is the outward unit normal vector at a point on the model boundary, and $\sigma = \eta \left(\nabla \mathbf{u} + \nabla \mathbf{u}^T \right)$ is the stress tensor. Zero heat flux is assumed through the vertical boundaries of the box. Either temperature or heat flux are prescribed at the upper and lower boundaries of the model domain. To solve the problem forward or backward in time, the temperature is considered to be known at the initial time ($t = 0$) or at the present time ($t = \vartheta$). Equations (3.1), (3.2) and (3.3) together with the boundary and initial conditions describe a thermal convective flow.

3.3 Objective Functional

Consider the following objective (cost) functional at $t \in [0, \vartheta]$

$$J (\varphi) = \| T (\vartheta, \cdot; \varphi) - \chi (\cdot) \|^2, \tag{3.4}$$

where $\|\cdot\|$ denotes the norm in the space $L_2(\Omega)$ (the Hilbert space with the norm defined as $\|y\| = \left[\int_\Omega y^2(\mathbf{x})\, d\mathbf{x}\right]^{1/2}$. Since in what follows the dependence of solutions of the thermal boundary value problems on initial data is important, these data are explicitly introduced into the mathematical representation of temperature. Here $T(\vartheta, \cdot; \varphi)$ is the solution of the thermal boundary value problem (3.1) at the final time ϑ, which corresponds to some (unknown as yet) initial temperature distribution $\phi(\mathbf{x})$; $\chi(\mathbf{x}) = T(\vartheta, \mathbf{x}; T_0)$ is the known temperature distribution at the final time, which corresponds to the initial temperature $T_0(\cdot)$. The functional has its unique global minimum at value $\varphi \equiv T_0$ and $J(T_0) \equiv 0$, $\nabla J(T_0) \equiv 0$ (Vasiliev 2002).

To find the minimum of the functional, the following gradient method is employed ($k = 0, \ldots, j, \ldots$):

$$\varphi_{k+1} = \varphi_k - \beta_k \nabla J(\varphi_k), \quad \varphi_0 = T_*, \tag{3.5}$$

$$\beta_k = \begin{cases} J(\varphi_k)/\|\nabla J(\varphi_k)\|^2, & 0 \le k \le k_* \\ k^{-1}, & k > k_* \end{cases}, \tag{3.6}$$

where T_* is an initial temperature guess. The minimization method belongs to a class of limited-memory quasi-Newton methods (Zou et al. 1993), where approximations to the inverse Hessian matrices are chosen to be the identity matrix. Equation (3.6) is used to maintain the stability of the iteration scheme (3.5).

Let us consider that the gradient of the objective functional $\nabla J(\varphi_k)$ is computed with an error $\|\nabla J_\delta(\varphi_k) - \nabla J(\varphi_k)\| < \delta$, where $\nabla J_\delta(\varphi_k)$ is the computed value of the gradient. Introducing the function $\varphi^\infty = \varphi_0 - \sum_{k=1}^\infty \beta_k \nabla J(\varphi_k)$ (and assuming that the infinite sum exists) and the function $\varphi_\delta^\infty = \varphi_0 - \sum_{k=1}^\infty \beta_k \nabla J_\delta(\varphi_k)$ (as the computed value of φ^∞), the following inequality should be held for stability of the iteration method (3.5):

$$\|\varphi_\delta^\infty - \varphi^\infty\| = \left\|\sum_{k=1}^\infty \beta_k (\nabla J_\delta(u_k) - \nabla J(u_k))\right\| \le \sum_{k=1}^\infty \beta_k \|\nabla J_\delta(\varphi_k) - \nabla J(\varphi_k)\|$$

$$\le \delta \sum_{k=1}^\infty \beta_k.$$

The sum $\sum_{k=1}^\infty \beta_k$ is finite, if $\beta_k = 1/k^p$, $p > 1$. If $p = 1$, but the number of iterations is limited, the iteration method is conditionally stable, although the convergence rate of these iterations is low. Meanwhile the gradient of the objective functional

$\nabla J\left(\varphi_{k}\right)$ decreases steadily with the number of iterations providing the convergence, although the absolute value of $J\left(\varphi_{k}\right)/\left\|\nabla J\left(\varphi_{k}\right)\right\|^{2}$ increases with the number of iterations, and it can result in instability of the iteration process (Samarskii and Vabishchevich 2007).

3.4   Adjoint Problem

The minimization algorithm requires the calculation of the gradient of the objective functional, ∇J. This can be done through the use of the *adjoint problem* for the model Eqs. (3.1), (3.2) and (3.3) with the relevant boundary and initial conditions. In the case of the heat problem, the adjoint problem can be represented in the following form:

$$\partial\Psi/\partial t + \langle\mathbf{u}, \nabla\Psi\rangle + \nabla^{2}\Psi = 0, \quad \mathbf{x}\in\Omega, \, t\in(0,\vartheta),$$

$$\sigma_{1}\Psi + \sigma_{2}\partial\Psi/\partial\mathbf{n} = 0, \quad \mathbf{x}\in\Gamma, \, t\in(0,\vartheta),$$

$$\Psi\left(\vartheta,\mathbf{x}\right) = 2\left(T\left(\vartheta,\mathbf{x};\varphi\right) - \chi\left(\mathbf{x}\right)\right), \quad \mathbf{x}\in\Omega, \tag{3.7}$$

where σ_{1} and σ_{2} are some smooth functions or constants satisfying the condition $\sigma_{1}^{2} + \sigma_{2}^{2} \neq 0$. Corresponding boundary conditions can be chosen by a selection of specific σ_{1} and σ_{2}.

The solution to the adjoint problem (3.7) is the gradient of the objective functional (3.4). To prove the statement, Ismail-Zadeh et al. (2004) considered an increment of the functional J in the following form:

$$J\left(\varphi + h\right) - J\left(\varphi\right) = \int_{\Omega}\left(T\left(\vartheta,\mathbf{x};\varphi + h\right) - \chi\left(\mathbf{x}\right)\right)^{2}d\mathbf{x} - \int_{\Omega}\left(T\left(\vartheta,\mathbf{x};\varphi\right) - \chi\left(\mathbf{x}\right)\right)^{2}d\mathbf{x}$$

$$= 2\int_{\Omega}\left(T\left(\vartheta,\mathbf{x};\varphi\right) - \chi\left(\mathbf{x}\right)\right)\zeta\left(\vartheta,\mathbf{x}\right)d\mathbf{x} + \int_{\Omega}\zeta^{2}\left(\vartheta,\mathbf{x}\right)d\mathbf{x}$$

$$= \int_{\Omega}\Psi\left(\vartheta,\mathbf{x}\right)\zeta\left(\vartheta,\mathbf{x}\right)d\mathbf{x} + \int_{\Omega}\zeta^{2}\left(\vartheta,\mathbf{x}\right)d\mathbf{x}$$

$$= \int_{\Omega}\int_{0}^{\vartheta}\frac{\partial}{\partial t}\left(\Psi\left(t,\mathbf{x}\right)\zeta\left(t,\mathbf{x}\right)\right)d\mathbf{x}dt + \int_{\Omega}\Psi\left(0,\mathbf{x}\right)h\left(\mathbf{x}\right)d\mathbf{x}$$

$$+ \int_{\Omega}\zeta^{2}\left(\vartheta,\mathbf{x}\right)d\mathbf{x}, \tag{3.8}$$

where $\Psi(t, \mathbf{x}) = 2(T(t, \mathbf{x}; \vartheta) - \chi(\mathbf{x}))$; $h(\mathbf{x})$ is a small heat increment to the unknown initial temperature $\varphi(\mathbf{x})$, and $\zeta = T(t, \mathbf{x}; \varphi + h) - T(t, \mathbf{x}; \varphi)$ is the solution to the following forward heat problem

$$\partial \zeta / \partial t + \langle \mathbf{u}, \nabla \zeta \rangle - \nabla^2 \zeta = 0, \quad \mathbf{x} \in \Omega, \ t \in (0, \vartheta),$$

$$\sigma_1 \zeta + \sigma_2 \partial \zeta / \partial \mathbf{n} = 0, \quad \mathbf{x} \in \Gamma, \ t \in (0, \vartheta),$$

$$\zeta(0, \mathbf{x}) = h(\mathbf{x}), \quad \mathbf{x} \in \Omega. \tag{3.9}$$

Considering the fact that $\Psi = \Psi(t, \mathbf{x})$ and $\zeta = \zeta(t, \mathbf{x})$ are the solutions to (3.7) and (3.9) respectively, and the velocity $\mathbf{u}$ satisfies (3.3) and the boundary conditions specified, the first term in (3.8) can be represented as

$$\int_\Omega \int_0^\vartheta \frac{\partial}{\partial t}(\Psi(t, \mathbf{x}) \zeta(t, \mathbf{x})) \, dt d\mathbf{x} = \int_0^\vartheta \int_\Omega \left\{ \frac{\partial}{\partial t} \Psi(t, \mathbf{x}) \zeta(t, \mathbf{x}) + \Psi(t, \mathbf{x}) \frac{\partial \zeta(t, \mathbf{x})}{\partial t} \right\} d\mathbf{x} dt$$

$$= \int_0^\vartheta \int_\Omega \zeta(t, x) \left[-\mathbf{u} \cdot \nabla \Psi - \nabla^2 \Psi \right] d\mathbf{x} dt + \int_0^\vartheta \int_\Omega \Psi(t, \mathbf{x}) \left[-\mathbf{u} \cdot \nabla \zeta + \nabla^2 \zeta \right] d\mathbf{x} dt$$

$$= \int_0^\vartheta \int_\Gamma \{ \Psi \nabla \zeta \cdot \mathbf{n} - \zeta \nabla \Psi \cdot \mathbf{n} \} d\Gamma dt + \int_0^\vartheta \int_\Omega \{ \nabla \Psi \cdot \nabla \zeta - \nabla \zeta \cdot \nabla \Psi \} d\mathbf{x} dt$$

$$+ \int_0^\vartheta \int_\Omega \{ \zeta \Psi \nabla \cdot \mathbf{u} + \Psi \mathbf{u} \cdot \nabla \zeta - \Psi \mathbf{u} \cdot \nabla \zeta \} \, d\mathbf{x} dt - 2 \int_0^\vartheta \int_\Gamma \zeta \Psi \mathbf{u} \cdot \mathbf{n} \, d\Gamma dt = 0.$$

$$\tag{3.10}$$

Hence

$$J(\varphi + h) - J(\varphi) = \int_\Omega \Psi(0, \mathbf{x}) h(\mathbf{x}) \, d\mathbf{x} + \int_\Omega \zeta^2(\vartheta, \mathbf{x}) \, d\mathbf{x} = \int_\Omega \Psi(0, \mathbf{x}) h(\mathbf{x}) \, d\mathbf{x} + o(\|h\|).$$

$$\tag{3.11}$$

The gradient is derived by using the Gateaux derivative of the objective functional. Therefore, the gradient of the functional is represented as $\nabla J(\varphi) = \Psi(0, \cdot)$. Thus, the solution of the backward heat problem is reduced to solutions of series of forward problems, which are known to be well-posed (Tikhonov and Samarskii 1990). The algorithm can be used to solve the problem over any subinterval of time in $[0, \vartheta]$.

3.5 Solution Method

Here the method for numerical solution of the inverse problem of thermal convection in the mantle is described. Namely, the numerical algorithm is presented to solve (3.1), (3.2) and (3.3) backward in time using the VAR method. A uniform partition of the time axis is defined at points $t_n = \vartheta - \delta t\, n$, where δt is the time step, and n successively takes integer values from 0 to some natural number $m = \vartheta/\delta t$. At each subinterval of time $[t_{n+1}, t_n]$, the search of the temperature T and flow velocity $\mathbf{u}$ at $t = t_{n+1}$ consists of the following basic steps.

Step 1. Given the temperature $T = T(t_n, \mathbf{x})$ at $t = t_n$ solve a set of linear algebraic equations derived from (3.2) and (3.3) with the appropriate boundary conditions in order to determine the velocity $\mathbf{u}$.

Step 2. The 'advective' temperature $T_{adv} = T_{adv}(t_{n+1}, \mathbf{x})$ is determined by solving the advection heat equation backward in time, neglecting the diffusion term in Eq. (3.1). This can be done by replacing positive time steps by negative ones. Given the temperature $T = T_{adv}$ at $t = t_{n+1}$ steps 1 and 2 are then repeated to find the velocity $\mathbf{u}_{adv} = \mathbf{u}(t_{n+1}, \mathbf{x}; T_{adv})$

Step 3. The heat Eq. (3.1) is solved with appropriate boundary conditions and initial condition $\varphi_k(\mathbf{x}) = T_{adv}(t_{n+1}, \mathbf{x})$, $k = 0, 1, 2, \ldots, m, \ldots$ forward in time using velocity $\mathbf{u}_{adv}$ in order to find $T(t_n, \mathbf{x}; \varphi_k)$.

Step 4. The adjoint equation of (3.7) is then solved backward in time with ppropriate boundary conditions and initial condition $\Psi(t_n, \mathbf{x}) = 2(T(t_n, \mathbf{x}; \varphi_k) - \chi(\mathbf{x}))$using velocity $\mathbf{u}$ in order to determine $\nabla J(\varphi_k) = \Psi(t_{n+1}, \mathbf{x}; \varphi_k)$.

Step 5. The coefficient β_k is determined from (3.6), and the temperature is updated (i.e. φ_{k+1} is determined) from (3.5).

Steps 3 to 5 are repeated until

$$\delta \varphi_n = J(\varphi_n) + \|\nabla J(\varphi_n)\|^2 < \varepsilon, \tag{3.12}$$

where ε is a small constant. Temperature φ_k is then considered to be the approximation to the target value of the initial temperature $T(t_{n+1}, \mathbf{x})$. And finally, step 1 is used to determine the flow velocity $\mathbf{u}(t_{n+1}, \mathbf{x}; T(t_{n+1}, \mathbf{x}))$. Step 2 introduces a pre-conditioner to accelerate the convergence of temperature iterations in Steps 3–5 at high Rayleigh number. At low Ra, Step 2 is omitted and $\mathbf{u}_{adv}$ is replaced by $\mathbf{u}$. After these algorithmic steps, temperature $T = T(t_n, \mathbf{x})$ and flow velocity $\mathbf{u} = \mathbf{u}(t_n, \mathbf{x})$ (corresponding to $t = t_n$, $n = 0, \ldots, m$) are obtained. Now based on the obtained results, and when required, interpolations can be used to reconstruct the process on the time interval $[0, \vartheta]$ in more detail.

Thus, at each subinterval of time:

– the VAR method is applied to the heat equation only;
– the direct and conjugate problems for the heat equation are solved iteratively to find temperature; and
– backward flow is determined from the Stokes and continuity equations twice (for 'advective' and 'true' temperatures).

Compared to the VAR approach by Bunge et al. (2003), the described numerical approach is computationally less expensive, because the Stokes equation is not involved into the iterations between the direct and conjugate problems (the numerical solution of the Stokes equation is the most time consuming calculation).

3.6 Restoration of Mantle Plumes

A plume is hot, narrow mantle upwelling that is invoked to explain hotspot volcanism. In a temperature-dependent viscosity fluid such as the mantle, a plume is characterized by a mushroom-shaped head and a thin tail. Upon impinging under a moving lithosphere, such a mantle upwelling should therefore produce a large amount of melt and successive massive eruption, followed by smaller but long-lived hot-spot activity fed from the plume tail (Morgan 1972; Richards et al. 1989; Sleep 1990). Meanwhile, slowly rising plumes (a buoyancy flux of less than 10^3 kg s^{-1}) coming from the core-mantle boundary should have cooled so much that they would not melt beneath old lithosphere (Albers and Christensen 1996).

Mantle plumes evolve in three distinguishing stages: *immature*, i.e. an origin and initial rise of the plumes; *mature*, i.e. plume-lithosphere interaction, gravity spreading of plume head and development of overhangs beneath the bottom of the lithosphere, and partial melting of the plume material (e.g. Ribe and Christensen 1994; Moore et al. 1998); and *overmature*, i.e. slowing-down of the plume rise and fading of the mantle plumes due to thermal diffusion (Davaille and Vatteville 2005; Ismail-Zadeh et al. 2006). The ascent and evolution of mantle plumes depend on the properties of the source region (that is, the thermal boundary layer) and the viscosity and thermal diffusivity of the ambient mantle. The properties of the source region determine temperature and viscosity of the mantle plumes. Structure, flow rate, and heat flux of the plumes are controlled by the properties of the mantle through which the plumes rise. While properties of the lower mantle (e.g. viscosity, thermal conductivity) are relatively constant during about 150 Myr lifetime of most plumes, source region properties can vary substantially with time as the thermal basal boundary layer feeding the plume is depleted of hot material (Schubert et al. 2001). Complete local depletion of this boundary layer cuts the plume off from its source.

A mantle plume is a well-established structure in computer modelling and laboratory experiments. Numerical experiments on dynamics of mantle plumes (Trompert and Hansen 1998; Zhong 2005) showed that the number of plumes increases and the rising plumes become thinner with an increase in Rayleigh number. Disconnected thermal plume structures appear in thermal convection at *Ra* greater than 10^7 (e.g., Hansen et al. 1990). At high *Ra* (in the hard turbulence regime) thermal plumes are torn off the boundary layer by the large-scale circulation or by nonlinear interactions between plumes (Malevsky and Yuen 1993). Plume tails can also be disconnected when the plumes are tilted by plate scale flow (e.g. Olson and Singer 1985). Ismail-Zadeh et al. (2006) presented an alternative explanation for the disconnected mantle plume heads and tails, which is based on thermal diffusion of mantle plumes.

3.6.1 Model and Methods

To model the evolution of mantle plumes, Ismail-Zadeh et al. (2006) used Eqs. (3.1), (3.2) and (3.3) with impenetrability and perfect slip conditions at the model boundary Ω (see Sect. 3.2). A temperature-dependent viscosity $\eta(T) = \exp\left(\frac{M}{T+G} - \frac{M}{0.5+G}\right)$ is employed, where $M = [225/\ln(r)] - 0.25 \ln(r)$, $G = 15/\ln(r) - 0.5$ and r is the viscosity ratio between the upper and lower boundaries of the model domain (Busse et al. 1993). The temperature-dependent viscosity profile has its minimum at the core-mantle boundary. A more realistic viscosity profile (e.g. Forte and Mitrovica 1997) will influence the evolution of mantle plumes, though it will not influence the restoration of the plumes.

The model domain is divided into $37 \times 37 \times 29$ rectangular finite elements to approximate the vector velocity potential by tricubic splines, and a uniform grid $112 \times 112 \times 88$ is employed for approximation of temperature, velocity, and viscosity (Ismail-Zadeh et al. 2006). Temperature in the heat Eq. (3.1) is approximated by finite differences and determined by the semi-Lagrangian method (e.g., Ismail-Zadeh and Tackley 2010; chapter 7.8). A numerical solution to the Stokes and incompressibility Eqs. (3.2 and 3.3) is based on the introduction of a two-component vector velocity potential and on the application of the Eulerian finite-element method with a tricubic-spline basis for computing the potential (e.g., Ismail-Zadeh and Tackley 2010; chapter 4.10.2). Such a procedure results in a set of linear algebraic equations solved by the conjugate gradient method (e.g., Ismail-Zadeh and Tackley 2010; chapter 6.3.3).

3.6.2 Forward Modelling

We present here the evolution of mature mantle plumes modelled forward in time. Considering the following model parameters, $\alpha = 3 \times 10^{-5}$ K^{-1}, $\rho_{ref} = 4000$ kg m^{-3}, $\Delta T = 3000$ K, $h = 2800$ km, $\eta_{ref} = 8 \times 10^{22}$ Pa s, and $\kappa = 10^{-6}$ m^{-2} s^{-1}, the Rayleigh number is estimated to be $Ra = 9.5 \times 10^{5}$. While plumes evolve in the convecting heterogeneous mantle, at the initial time it is assumed that the plumes develop in a laterally homogeneous temperature field, and hence the initial mantle temperature is considered to increase linearly with depth.

Mantle plumes are generated by random temperature perturbations at the top of the thermal source layer associated with the core-mantle boundary (Fig. 3.1a). The mantle material in the basal source layer flows horizontally toward the plumes. The reduced viscosity in this basal layer promotes the flow of the material to the plumes. Vertical upwelling of hot mantle material is concentrated in low viscosity conduits near the centrelines of the emerging plumes (Fig. 3.1b, c). The plumes move upward through the model domain, gradually forming structures with well-developed heads and tails. Colder material overlying the source layer (e.g. portions of lithospheric slabs subducted to the core-mantle boundary) replaces hot material at the locations where the source material is fed into mantle plumes. Some time is required to recover the volume of source material depleted due to plume feeding

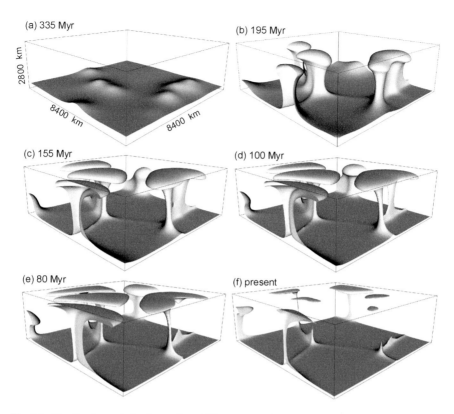

Fig. 3.1 Mantle plumes in the forward modelling at successive diffusion times: from 335 Myr ago to the 'present' state of the plumes. The plumes are represented here and in Fig. 3.2 by isothermal surfaces at 3000 K. (After Ismail-Zadeh et al. 2006)

(Howard 1966). Because the volume of upwelling material is comparable to the volume of the thermal source layer feeding the mantle plumes, hot material could eventually be exhausted, and mantle plumes would be starved thereafter.

The plumes diminish in size with time (Fig. 3.1d), and the plume tails disappear before the plume heads (Fig. 3.1e, f). Figure 3.1 presents a hot isothermal surface of the plumes; if colder isotherms are considered, the disappearance of the isotherms will occur later. But anyhow, hot or cold isotherms are plotted, plume tails will vanish before their heads. Results of laboratory experiments (Davaille and Vatteville 2005) support the numerical findings by Ismail-Zadeh et al. (2006), presented here, that plumes start disappearing from bottom up and fade away by thermal diffusion.

At different stages in the plume decay one sees quite isolated plume heads, plume heads with short tails, and plumes with nearly pinched off tails. Different amounts of time are required for different mantle plumes to vanish into the ambient mantle, the required time depending on the geometry of the plume tails. Temperature loss is greater for sheet-like tails than for cylindrical tails. The tails of the cylindrical plumes (e.g. Fig. 3.1c, in the left part of the model domain) are still detectable after about 155 Myr. However, at this time the sheet-like tail of the large plume in the right part of the model domain (Fig. 3.1c) is already invisible and only its head is preserved in the uppermost mantle (Fig. 3.1f).

3.6.3 Backward Modelling

To restore the prominent state of the plumes (Fig. 3.1d) in the past from their 'present' weak state (Fig. 3.1f), Ismail-Zadeh et al. (2006) employed the VAR method. Figure 3.2 illustrates the restored states of the plumes (middle panel) and the temperature residuals δT (right panel) between the temperature $T(\mathbf{x})$ predicted by the forward model and the temperature $\tilde{T}(\mathbf{x})$ reconstructed to the same age:

$$\delta T (x_1, x_2) = \left[\int_0^h \left(T(x_1, x_2, x_3) - \tilde{T}(x_1, x_2, x_3) \right)^2 dx_3 \right]^{1/2}. \qquad (3.13)$$

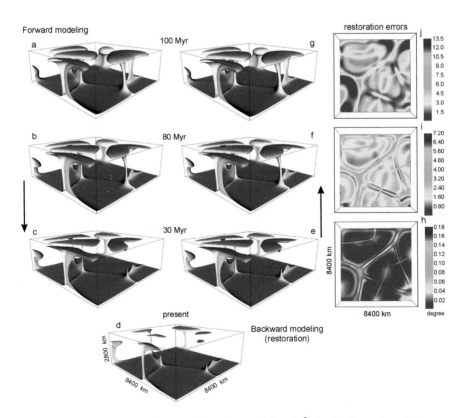

Fig. 3.2 Mantle plume diffusion ($r = 20$ and $Ra = 9.5 \times 10^5$) in the forward modelling at successive diffusion times: from 100 Myr ago to the 'present' state of the plumes (*left panel*, **a–d**). Restored mantle plumes in the backward modelling (*central panel*, **e–g**) and restoration errors (*right panel*, **h–j**) (After Ismail-Zadeh et al. 2006)

To study the effect of thermal diffusion on the restoration of mantle plumes, Ismail-Zadeh et al. (2006) performed several experiments on mantle plume restoration for various Rayleigh number Ra and viscosity ratio r. The dimensional temperature residuals are within a few degrees for the initial restoration period (Fig. 3.2h, i). The computations show that the errors (temperature residuals) get larger the farther the restorations move backward in time (e.g. $\delta T \approx 300$ K at the restoration time of more than 300 Myr, $r = 200$, and $Ra = 9.5 \times 10^3$). One can see that the residuals become larger as the Rayleigh number decreases or thermal diffusion increases and viscosity ratio increases.

The quality of the restoration depends on the dimensionless Peclet number $Pe = hu_{max}\kappa^{-1}$, where u_{max} is the maximum flow velocity. According to the numerical experiments, the Peclet number corresponding to the temperature residual $\delta T = 600$ K is $Pe = 10$; Pe should not be less than about 10 for a high quality plume restoration.

3.6.4   Performance of the Numerical Algorithm

The performance of the algorithm for various Ra and r is evaluated in terms of the number of iterations n required to achieve a prescribed relative reduction of $\delta\phi_n$ (inequality 3.12). Figure 3.3 presents the evolution of the objective functional $J(\phi_n)$ and the norm of the gradient of the objective functional $\|\nabla J (\varphi_n)\|$ versus the number of iterations at time about 0.5θ. For other time steps a similar evolution of J and $\|\nabla J\|$ is found.

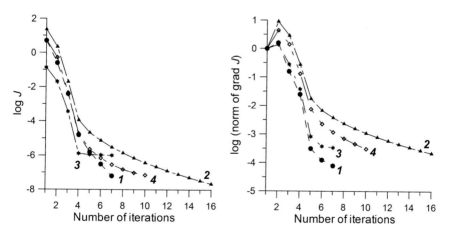

Fig. 3.3 Relative reductions of the objective functional J (*left panel*) and the norm of the gradient of J (*right panel*) as functions of the number of iterations. Curves: *1*, $r = 20$, $Ra = 9.5 \times 10^5$; *2*, $r = 20$, $Ra = 9.5 \times 10^2$; *3*, $r = 200$, $Ra = 9.5 \times 10^3$; *4*, $r = 200$, $Ra = 9.5 \times 10^2$ (After Ismail-Zadeh et al. 2006)

Both the objective functional and the norm of its gradient show a quite rapid decrease after about 7 iterations for $Ra = 9.5 \times 10^5$ and $r = 20$ (curves 1). As Ra decreases and thermal diffusion increases (curves 2–4) the performance of the algorithm becomes poor: more iterations are needed to achieve the prescribed ε. All curves illustrate that the first 4–7 iterations contribute mainly to the reduction of $\delta\phi_n$. The convergence drops after a relatively small number of iterations. The curves approach the horizontal line with an increase in the number of iterations, because β_k tends to zero with a large number of iterations (see Eq. 3.6). The increase of $\|\nabla J\|$ at $k = 2$ is associated with uncertainty of this gradient at $k = 1$.

Implementation of minimization algorithms requires the evaluation of both the objective functional and its gradient. Each evaluation of the objective functional requires an integration of the model Eq. (3.1) with the appropriate boundary and initial conditions, whereas the gradient is obtained through the backward integration of the adjoint Eq. (3.7). The performance analysis shows that the CPU time required to evaluate the gradient J is about the CPU time required to evaluate the objective functional itself, and this is because the direct and adjoint heat problems are described by the same equations.

Despite its simplicity, the minimization algorithm (3.5) provides for a rapid convergence and good quality of optimisation at high Rayleigh numbers (Ismail-Zadeh et al. 2006). The convergence rate and the quality of optimisation become worse with the decreasing Rayleigh number. The use of the limited-memory quasi-Newton algorithm L-BFGS (Liu and Nocedal 1989) might provide for a better convergence rate and quality of optimisation (Zou et al. 1993; see a comparison of the methods in Chap. 4). Although an improvement of the convergence rate by using another minimization algorithm (e.g. L-BFGS) will reduce the computational expense associated with the solving of the problem under question, this reduction would be not significant, because the large portion (about 70 %) of the computer time is spent to solve the 3-D Stokes equations.

3.7 Challenges in VAR Data Assimilation

The VAR method for data assimilation can theoretically be applied to many geodynamic problems, although a practical implementation of the technique for modelling of real geodynamic processes backward in time is not a simple task. The mathematical model of mantle dynamics described by a set of Eqs. (3.1), (3.2) and (3.3) is simple, and many complications are omitted. For example, in the considered case study a viscosity increase from the upper to the lower mantle is not included in the model, although it is suggested by studies of the geoid (Ricard et al. 1984), post-glacial rebound (Mitrovica 1996), and joint inversion of convection and glacial isostatic adjustment data (Mitrovica and Forte 2004). The adiabatic heating/cooling term in the heat equation can provide more realistic distribution of temperature in the mantle, especially near the thermal boundary layer. The numerical models considered here do not include phase transformations (e.g. Liu et al. 1991; Honda et al. 1993a, b; Harder and Christensen 1996), although the phase changes can influence the thermal convection pattern. The coefficient of thermal expansion (e.g. Chopelas and Boehler 1989; Hansen et al. 1991) and the coefficient of thermal conductivity (e.g. Hofmeister 1999) are not constant in the mantle and vary with depth and temperature. To consider these complications in the VAR data assimilation, the adjoint equations should be derived each time when the set of the equations is changed. The cost to be paid is in software development since an adjoint model has to be developed.

3.7.1 Data Smoothness

The solution of the heat Eq. (3.1) with appropriate boundary and initial conditions is a sufficiently smooth function. The temperature derived from the seismic tomography is a representation of the exact temperature in the Earth interior, and so it must be rather smooth, because, otherwise, the objective functional cannot be defined. Therefore, before any assimilation of the present temperature data can be attempted, the data must be smoothed. The smoothing of the present temperature improves the convergence of the iterations.

  If the initial temperature is not a smooth function of space variables, recovery of this temperature using the VAR method is not effective because the iterations converge very slowly to the target temperature. Ismail-Zadeh et al. (2006) explained the problem of recovering the initial temperature on the basis of three one-dimensional model tasks: restoration of a (i) smooth, (ii) piece-wise smooth, and (iii) discontinuous target function. (We note that the temperature in the Earth's mantle is not a discontinuous function but its shape can be close to a step function.)

  Consider the dynamics of a physical system described by the Burgers equation (Ismail-Zadeh et al. 2006)

$$u_t + uu_x = u_{xx}, \ \ 0 \le t \le 1, \ 0 \le x \le 2\pi$$

with the boundary conditions

$$u(t,0) = 0, \ u(t,2\pi) = 0, \ 0 \le t \le 1,$$

and the condition

$$u_\theta = u(1,x;u_0), \ \ 0 \le x \le 2\pi \text{ at } t = 1,$$

where the variable u may denote temperature. The problem is to recover the function $u_0 = u_0(x), \ 0 \le x \le 2\pi$ at $t = 0$ (the state in the past) from the function $u_\theta = u_\theta(x), \ 0 \le x \le 2\pi$ at $t = 1$ (its present state). The finite difference approximations and the variational method are applied to the Burgers equation with the appropriate boundary and initial conditions.

Task 1. Consider the sufficiently smooth function $u_0 = \sin(x)$, $0 \le x \le 2\pi$. The functions u_0 and u_θ are shown in Fig. 3.4a. Figure 3.4b, c illustrate the iterations ϕ_k using the iterative scheme similar to (3.5) for $k = 0, 4, 6$ and the residual $r_6(x) = u_0(x) - \varphi_6(x)$, $0 \le x \le 2\pi$ respectively. Iterations converge rather rapid for the sufficiently smooth target function.

Task 2. Now consider the continuous piece-wise smooth function $u_0 = 3x/(2\pi)$, $0 \le x \le 2\pi/3$ and $u_0 = 3/2 - 3x/(2\pi)$, $2\pi/3 \le x \le 2\pi$. Figure 3.4 presents (d) the functions u_0 and u_θ, (e) the successive approximations ϕ_k for $k = 0, 4, 1000$, and (f) the residual $r_{1000}(x) = u_0(x) - \varphi_{1000}(x)$, $0 \le x \le 2\pi$, respectively. This example shows that a large number of iterations is required to reach the target function.

Task 3. Consider the discontinuous function u_0, which takes 1 at $2\pi/3 \le x \le 4\pi/3$ and 0 in other points of the closed interval $0 \le x \le 2\pi$. Figure 3.4 presents (g) the functions u_0 and u_θ, (h) the successive approximations ϕ_k for $k = 0, 500, 1000$, and (e) the residual $r_{1000}(x) = u_0(x) - \varphi_{1000}(x)$, $0 \le x \le 2\pi$, respectively. Convergence to the target temperature is very poor.

To improve the convergence to the target function, a modification of the variational method based on a priori information about a desired solution can be used (e.g., the maximum and minimum of the solution; Korotkii and Tsepelev 2003). Figure 3.4j shows the successive approximations $\tilde{\varphi}_k$ for $k = 0, 30, 500$, and (k) the residual $\tilde{r}_{500}(x) = u_0(x) - \tilde{\varphi}_{500}(x)$, $0 \le x \le 2\pi$, respectively. The approximations $\tilde{\varphi}_k$ based on the method of gradient projection (Vasiliev 2002) converge to the target solution better than approximations generated by (3.5).

3.7.2 Numerical Noise

If the initial temperature guess ϕ_0 is a smooth function, all successive temperature iterations ϕ_k in scheme (3.5) should be smooth functions too, because the gradient of the objective functional ∇J is a smooth function as it is the solution to the adjoint problem (3.7). However, the temperature iterations ϕ_k are polluted by small perturbations (errors), which are inherent in any numerical experiment. These perturbations can grow with time. Samarskii et al. (1997) applied a VAR method to a 1-D backward heat diffusion problem and showed that the solution to this problem becomes noisy if the initial temperature guess is slightly perturbed, and the amplitude of this noise increases with the initial perturbations of the temperature guess. To reduce the noise they used a special filter and illustrated the efficiency of the filter. This filter is based on the replacement of iterations (3.5) by $\mathbf{B}(\varphi_{k+1} - \varphi_k) = -\beta_k \nabla J(\varphi_k)$, where $\mathbf{B}y = y - \nabla^2 y$ (Tsepelev 2011). An employment of this filter increases the number of iterations to obtain the target temperature, and it becomes quite expensive computationally, especially when the model is three-dimensional. Another way to reduce the noise is to employ high-order adjoint (Alekseev and Navon 2001) or regularization (e.g. Tikhonov 1963; Lattes and Lions 1969, Samarskii and Vabischevich 2007) techniques.

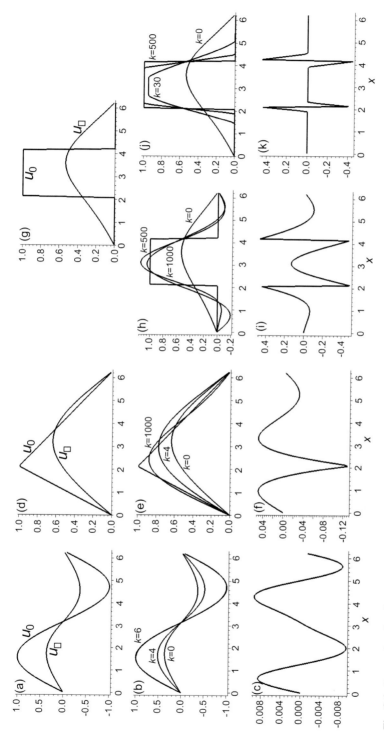

Fig. 3.4 Recovering function u_0 from the smooth guess function u_θ. The sufficiently smooth u_0 (**a–c**); continuous piece-wise smooth function u_0 (**d–f**); and discontinuous function u_0 (**g–k**). Plots of u_0 and u_θ are presented at (**a**), (**d**), and (**g**); successive approximations to u_0 at (**b**), (**e**), (**h**), and (**j**); and the residual functions at (**c**), (**f**), (**i**), and (**k**) (After Ismail-Zadeh et al. 2006)

References

Albers M, Christensen UR (1996) The excess temperature of plumes rising from the core-mantle boundary. Geophys Res Lett 23:3567–3570

Alekseev AK, Navon IM (2001) The analysis of an ill-posed problem using multiscale resolution and second order adjoint techniques. Comput Meth Appl Mech Eng 190:1937–1953

Boussinesq J (1903) Theorie Analytique de la Chaleur, vol 2. Gauthier-Villars, Paris

Bunge H-P, Hagelberg CR, Travis BJ (2003) Mantle circulation models with variational data assimilation: inferring past mantle flow and structure from plate motion histories and seismic tomography. Geophys J Int 152:280–301

Busse FH, Christensen U, Clever R, Cserepes L, Gable C, Giannandrea E, Guillou L, Houseman G, Nataf H-C, Ogawa M, Parmentier M, Sotin C, Travis B (1993) 3D convection at infinite Prandtl number in Cartesian geometry – a benchmark comparison. Geophys Astrophys Fluid Dyn 75:39–59

Chandrasekhar S (1961) Hydrodynamic and hydromagnetic stability. Oxford University Press, Oxford

Chopelas A, Boehler R (1989) Thermal expansion measurements at very high pressure, systematics and a case for a chemically homogeneous mantle. Geophys Res Lett 16:1347–1350

Davaille A, Vatteville J (2005) On the transient nature of mantle plumes. Geophys Res Lett 32:L14309. doi:10.1029/2005GL023029

Forte AM, Mitrovica JX (1997) A resonance in the Earth's obliquity and precession over the past 20 Myr driven by mantle convection. Nature 390:676–680

Hansen U, Yuen DA, Kroening SE (1990) Transition to hard turbulence in thermal convection at infinite Prandtl number. Phys Fluids A2(12):2157–2163

Hansen U, Yuen DA, Kroening SE (1991) Effects of depth-dependent thermal expansivity on mantle circulations and lateral thermal anomalies. Geophys Res Lett 18:1261–1264

Harder H, Christensen UR (1996) A one-plume model of Martian mantle convection. Nature 380:507–509

Hofmeister AM (1999) Mantle values of thermal conductivity and the geotherm from photon lifetimes. Science 283:1699–1706

Honda S, Balachandar S, Yuen DA, Reuteler D (1993a) Three-dimensional mantle dynamics with an endothermic phase transition. Geophys Res Lett 20:221–224

Honda S, Yuen DA, Balachandar S, Reuteler D (1993b) Three-dimensional instabilities of mantle convection with multiple phase transitions. Science 259:1308–1311

Howard LN (1966) Convection at high Rayleigh number. In: Goertler H, Sorger P (eds) Applied mechanics. In: Proceedings of the 11th international congress of applied mechanics, Munich, Germany 1964. Springer, New York, pp 1109–1115

Ismail-Zadeh A, Tackley P (2010) Computational methods for geodynamics. Cambridge University Press, Cambridge

Ismail-Zadeh AT, Korotkii AI, Tsepelev IA (2003) Numerical approach to solving problems of slow viscous flow backwards in time. In: Bathe KJ (ed) Computational fluid and solid mechanics. Elsevier Science, Amsterdam, pp 938–941

Ismail-Zadeh A, Schubert G, Tsepelev I, Korotkii A (2004) Inverse problem of thermal convection: numerical approach and application to mantle plume restoration. Phys Earth Planet Inter 145:99–114

Ismail-Zadeh A, Schubert G, Tsepelev I, Korotkii A (2006) Three-dimensional forward and backward numerical modeling of mantle plume evolution: effects of thermal diffusion. J Geophys Res 111:B06401. doi:10.1029/2005JB003782

Karato S (2010) Rheology of the Earth's mantle: a historical review. Gondwana Res 18:17–45

Korotkii AI, Tsepelev IA (2003) Solution of a retrospective inverse problem for one nonlinear evolutionary model. Proc Steklov Inst Math 243(Suppl 2):80–94

Lattes R, Lions JL (1969) The method of quasi-reversibility: applications to partial differential equations. Elsevier, New York

Liu DC, Nocedal J (1989) On the limited memory BFGS method for large scale optimization. Math Program 45:503–528

Liu M, Yuen DA, Zhao W, Honda S (1991) Development of diapiric structures in the upper mantle due to phase transitions. Science 252:1836–1839

Malevsky AV, Yuen DA (1993) Plume structures in the hard-turbulent regime of three-dimensional infinite Prandtl number convection. Geophys Res Lett 20:383–386

Mitrovica JX (1996) Haskell (1935) revisited. J Geophys Res 101:555–569

Mitrovica JX, Forte AM (2004) A new inference of mantle viscosity based upon joint inversion of convection and glacial isostatic adjustment data. Earth Planet Sci Lett 225:177–189

Moore WB, Schubert G, Tackley P (1998) Three-dimensional simulations of plume–lithosphere interaction at the Hawaiian Swell. Science 279:1008–1011

Morgan WJ (1972) Plate motions and deep convection. Geol Soc Am Mem 132:7–22

Olson P, Singer H (1985) Creeping plumes. J Fluid Mech 158:511–531

Ribe NM, Christensen U (1994) Three-dimensional modeling of plume-lithosphere interaction. J Geophys Res 99:669–682

Ricard Y, Fleitout L, Froidevaux C (1984) Geoid heights and lithospheric stresses for a dynamic arth. Ann Geophys 2:267–286

Richards MA, Duncan RA, Courtillot V (1989) Flood basalts and hot spot tracks: plume heads and tails. Science 246:103–107

Samarskii AA, Vabishchevich PN (2007) Numerical methods for solving inverse problems of mathematical physics. De Gruyter, Berlin

Samarskii AA, Vabishchevich PN, Vasiliev VI (1997) Iterative solution of a retrospective inverse problem of heat conduction. Math Modeling 9:119–127

Schubert G, Turcotte DL, Olson P (2001) Mantle convection in the earth and planets. Cambridge University Press, Cambridge

Sleep NH (1990) Hotspots and mantle plumes: some phenomenology. J Geophys Res 95:6715–6736

Tikhonov AN (1963) Solution of incorrectly formulated problems and the regularization method. Dokl Akad Nauk SSSR 151:501–504 (Engl. transl.: Soviet Math Dokl 4:1035–1038)

Tikhonov AN, Samarskii AA (1990) Equations of mathematical physics. Dover Publications, New York

Trompert RA, Hansen U (1998) On the Rayleigh number dependence of convection with a strongly temperature-dependent viscosity. Phys Fluids 10:351–360

Tsepelev IA (2011) Iterative algorithm for solving the retrospective problem of thermal convection in a viscous fluid. Fluid Dyn 46:835–842

Vasiliev FP (2002) Methody optimizatsii. Factorial Press, Moscow (in Russian)

Zhong S (2005) Dynamics of thermal plumes in three-dimensional isoviscous thermal convection. Geophys J Int 162:289–300

Zou X, Navon IM, Berger M, Phua KH, Schlick T, Le Dimet FX (1993) Numerical experience with limited-memory quasi-Newton and truncated Newton methods. SIAM J Optimi 3(3):582–608

Chapter 4
Application of the Variational Method to Lava Flow Modelling

Abstract In this chapter, we present an application of the variational data assimilation method to the problem for determination of thermal and dynamic characteristics of lava flow from thermal measurements at lava's upper surface. Assuming that the temperature and the heat flow are known at the lava's upper surface, the missing condition at the lower surface of the lava is determined at first, and then the flow characteristics (temperature and flow velocity) are resolved in the entire model domain.

Keywords Lava flow • Inverse theory • Adjoint problem • VAR method • Numerical modelling

4.1 Lava Flow

During volcanic non-explosive (effusive) eruptions a lava flow starts to form when partially molten rock is erupted onto the Earth's surface and spreads slowly on the surface from the volcanic edifice. The eruptions produce a variety of gravity currents depending on temperature and the chemical composition of the magmatic rocks, and the topography of the surface over which the lava flows (Griffiths 2000). Under relatively steady eruption conditions, a viscous lava flow rapidly forms a solid crust. A large surface heat flux from the lava, cooling, and crystallization of the uppermost layer of the moving melt lead to a gravity current of lava under a solid crust, which insulates the lava flow interior. The crust preserves the lava against rapid cooling and permits the lava flow extending to substantial distances. Once the lava supply ceases and the interior of the lava flow cools, the lava stops its further advance (Harris et al. 2007).

Computer simulations play an important role in understanding the dynamics, the morphology and thermal structures of lava flows (e.g., Costa and Macedonio 2005a and references herein). Simplified isothermal models of viscous flows have demonstrated the way, in which slow eruptions of lava would advance in the absence of cooling. These provided the basis for models that include heterogeneous rheology and change caused by cooling and solidification. A more realistic approach is to calculate mass and energy transport in viscous flow using a digital elevation

© The Author(s) 2016
A. Ismail-Zadeh et al., *Data-Driven Numerical Modelling in Geodynamics:
Methods and Applications*, SpringerBriefs in Earth Sciences,
DOI 10.1007/978-3-319-27801-8_4

model to represent topography. A 2-D model for lava flow was developed by Ishihara et al. (1989) and later refined by Miyamoto and Sasaki (1998). Another 2-D model, which was based on the conservation equations for lava thickness and depth-averaged velocities and temperature, was developed by Costa and Macedonio (2005b). Solidification occurs during lava spreading and consequently the solidified crust can become obstacles to the lava flow. Hidaka et al. (2005) developed the lava flow simulation code based on 3-D convection analysis with simultaneous spreading and solidification. Tsepelev et al. (2016) developed numerical models of fluid flow with breccia for various scenarios of lava advancement.

4.2 Reconstruction of Lava Properties

Modern remote sensing technologies (e.g., air-borne or space-borne infrared sensors) allow for detecting the absolute temperature at the Earth's surface (e.g., Flynn et al. 2001). The Stefan-Boltzmann law relates the total energy radiated per unit surface area of a body across all wavelengths per unit time to the fourth power of the absolute temperature of the body. Hence the absolute temperature can be determined from the measurements by remote sensors (e.g., Harris et al. 2004). The heat flow could be then inferred from the Stefan-Boltzmann law using the temperature.

Is it possible to use the surface thermal data so obtained to constrain the thermal and dynamic conditions beneath the surface? Following Korotkii et al. (2016) we present in this chapter a quantitative approach to reconstruct temperature and velocity in the steady-state lava flow. The knowledge of the thermal and dynamic characteristics of lava is important, particularly, for lava flow hazard assessment and hence disaster mitigation (Cutter et al. 2015).

The problem of reconstruction of lava thermal and flow characteristics is considered in the case when the temperature and the heat flow are known on the lava surface, but the lava temperature and velocity are unknown. The problem is reduced to determination of temperature and velocity as the solution to the model of steady-state flow of viscous heterogeneous incompressible fluid with prescribed conditions for velocity and temperature at the boundary $\Gamma = \partial\Omega$ of the model domain Ω. At a part of the model boundary the conditions are abundant (e.g. both temperature and heat flow are known), and at another part of the boundary there is a lack of information on the temperature (because of no direct measurements at this part of the boundary). This mathematical problem is reduced in its turn to solving the inverse problem for determination of the temperature at the bottom of the lava and for subsequent search for the temperature and velocity of the lava.

4.3 Mathematical Statement

In a two-dimensional model domain Ω (Fig. 4.1) the Stokes, incompressibility and heat equations are employed to determine the steady-state flow velocity and temperature of the incompressible heterogeneous fluid under gravity in the Boussinesq approximation:

$$\nabla \cdot \left(\eta \left(\nabla \mathbf{u} + \nabla \mathbf{u}^T \right) \right) = \nabla p - Ra\, T\, \mathbf{e}_2, \qquad (4.1)$$

$$\nabla \cdot \mathbf{u} = 0, \qquad (4.2)$$

$$\nabla \cdot (\kappa \, \nabla T) = \langle \mathbf{u}, \nabla T \rangle, \qquad (4.3)$$

where $\mathbf{x} = (x_1, x_2) \in \Omega$ are the Cartesian coordinates; $\mathbf{u} = (u_1(\mathbf{x}), u_2(\mathbf{x}))$ is the vector velocity; $p = p(\mathbf{x})$ is the pressure; $T = T(\mathbf{x})$ is the temperature; $\eta = \eta(T)$ is the viscosity; $\kappa = k/(\rho_{ref} c_p)$ is the thermal diffusivity; $k = k(T)$ is the heat conductivity; ρ_{ref} is the typical density; and c_p is the specific heat capacity. The Rayleigh number is defined as $Ra = \alpha g \rho_{ref} \Delta T h^3 \mu_{ref}^{-1} \kappa_{ref}^{-1}$, where α is the thermal expansivity; g is the acceleration due to gravity; η_{ref} and κ_{ref} are the typical viscosity and thermal diffusivity, respectively; ΔT is the temperature contrast; h is the typical length; $\mathbf{e}_2 = (0, -1)$ is the unit vector; ∇, ^{T}, and $\langle \cdot, \cdot \rangle$ denote the gradient vector, the transposed matrix, and the scalar product of vectors, respectively. Length and temperature are normalized by h and ΔT, respectively.

The following conditions for temperature and velocity are assumed at the model boundary $\Gamma = \Gamma_1 \cup \Gamma_2 \cup \Gamma_3 \cup \Gamma_4$. The temperature T_1 and the velocity $\mathbf{u}_1$ are prescribed at the left boundary Γ_1:

Fig. 4.1 Geometry of the lava flow model

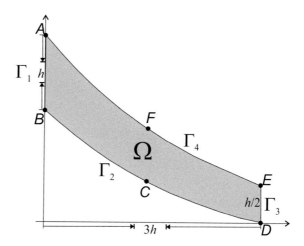

$$T = T_1, \quad \mathbf{u} = \mathbf{u}_1. \tag{4.4}$$

No slip condition is prescribed at the lower boundary Γ_2 (unknown temperature is to be found):

$$\mathbf{u} = 0. \tag{4.5}$$

At the right boundary Γ_3, the temperature T_3 is prescribed (a strong assumption, which might be omitted, but will complicate the problem solution), the deviatoric stress and the pressure are vanishing:

$$T = T_3, \quad \sigma \, \mathbf{n} = 0, \quad p = 0. \tag{4.6}$$

At the upper surface Γ_4, the temperature T_4 and heat flow φ are given, and no normal flow and free-slip tangential conditions are used:

$$T = T_4, \quad k \langle \nabla T, \mathbf{n} \rangle = \varphi, \quad \langle \mathbf{u}, \mathbf{n} \rangle = 0, \quad \sigma \mathbf{n} - \langle \sigma \, \mathbf{n}, \mathbf{n} \rangle \, \mathbf{n} = 0, \tag{4.7}$$

where $\sigma = \eta \left(\nabla \mathbf{u} + \nabla \mathbf{u}^T \right)$ is the deviatoric stress tensor, and $\mathbf{n}$ is the outward unit normal vector at a point on the model boundary. The principal problem is to find the solution to Eqs. (4.1), (4.2) and (4.3) with the boundary conditions (4.4), (4.5), (4.6) and (4.7), and hence to determine the velocity $\mathbf{u} = \mathbf{u}(\mathbf{x})$, the pressure $p = p(\mathbf{x})$, and the temperature $T = T(\mathbf{x})$ in the model domain Ω when temperature T_4 and heat flow $\varphi = k \partial T / \partial \mathbf{n}$ are known at boundary Γ_4.

In addition to the principal problem, an auxiliary problem is defined as: to find solution to Eqs. (4.1), (4.2) and (4.3) (that is, to determine $\mathbf{u}$, p, and T in Ω) with the following boundary conditions:

$$\Gamma_1 : \quad T = T_1, \quad \mathbf{u} = \mathbf{u}_1, \tag{4.8}$$

$$\Gamma_2 : \quad T = T_2, \quad \mathbf{u} = 0, \tag{4.9}$$

$$\Gamma_3 : \quad T = T_3, \quad \sigma \, \mathbf{n} = 0, \quad p = 0, \tag{4.10}$$

$$\Gamma_4 : \quad T = T_4, \quad \langle \mathbf{u}, \mathbf{n} \rangle = 0, \quad \sigma \mathbf{n} - \langle \sigma \, \mathbf{n}, \mathbf{n} \rangle \, \mathbf{n} = 0. \tag{4.11}$$

The auxiliary problem (4.1), (4.2) and (4.3) and (4.8), (4.9), (4.10) and (4.11) is a direct problem compared to the problem (4.1), (4.2), (4.3), (4.4), (4.5), (4.6) and (4.7), which is an inverse problem. The conditions at Γ_1 and Γ_3 are the same in the direct and inverse problems, but the temperature T_2 is known at Γ_2 and no heat flow is prescribed at Γ_4 in the auxiliary problem compared to the inverse problem (4.1), (4.2), (4.3), (4.4), (4.5), (4.6) and (4.7). The well- and ill-posedness of the similar problems have been studied by Ladyzhenskaya (1969), Lions (1971), Temam (1977), Korotkii and Kovtunov (2006), and Korotkii and Starodubtseva (2014).

The (measured) heat flow $\varphi = k(T)\partial T/\partial \mathbf{n}$ at model boundary Γ_4 is assumed to be related to some (unknown as yet) temperature $T = T_2 = \xi*$ at model boundary Γ_2, and temperature $T*$ is a component of the solution $(T*, \mathbf{u}*, p*)$ to the auxiliary problem, when the temperature $T = T_2$ at Γ_2 equals to $\xi*$ (Eq. 4.9), and hence $\varphi = k(T*)\partial T*/\partial \mathbf{n}$ at Γ_4.

Consider the cost functional for admissible functions ξ determined at Γ_2

$$J(\xi) = \int_{\Gamma_4} \left(k\left(T_\xi\right) \frac{\partial T_\xi}{\partial \mathbf{n}} - \varphi \right)^2 d\Gamma, \tag{4.12}$$

where T_ξ is the component of the solution $(T_\xi, \mathbf{u}_\xi, p_\xi)$ of the auxiliary problem with the condition $T = \xi$ at Γ_2 in Eq. (4.9). The functional has its global minimum at value $\xi = \xi*$ and $J(\xi*) = 0$, that is, temperature $\xi = \xi*$ attains a minimal value to the functional

$$J(\xi) \to \min : \xi \in \Xi, \tag{4.13}$$

where Ξ denotes a set of admissible temperatures at boundary Γ_2. Therefore, the inverse problem is reduced to a minimization of the functional or to a variation of the function ξ at Γ_2, so that heat flow $k\partial T/\partial \mathbf{n}$ at Γ_4 becomes closer to the prescribed value φ at Γ_4.

4.4 Minimisation Problem

To minimise the cost functional (4.12) the Polak-Ribière conjugate-gradient method is employed (e.g., Polak 1997):

$$\xi^{(n+1)} = \xi^{(n)} + \gamma^{(n)} d^{(n)}, \quad n = 1, 2, 3, \ldots, \tag{4.14}$$

$$d^{(n)} = \begin{cases} -\nabla J\left(\xi^{(n)}\right), & n = 1 \\ -\nabla J\left(\xi^{(n)}\right) + \beta^{(n)} d^{(n-1)}, & n = 2, 3, \ldots \end{cases}, \tag{4.15}$$

$$\beta^{(n)} = \int_{\Gamma_2} \nabla J\left(\xi^{(n)}\right)\left(\nabla J(\xi^{(n)}) - \nabla J\left(\xi^{(n-1)}\right)\right) d\Gamma / \int_{\Gamma_2} \left(\nabla J\left(\xi^{(n-1)}\right)\right)^2 d\Gamma, \quad n = 2, 3, \ldots, \tag{4.16}$$

and the descent step length $\gamma^{(n)}$ can be found from the Wolfe conditions (e.g., Nocedal and Wright 1999):

$$\begin{cases} J\left(\xi^{(n)} + \gamma^{(n)}d^{(n)}\right) \leq J\left(\xi^{(n)}\right) + c_1\,\gamma^{(n)}\displaystyle\int_{\Gamma_2} \nabla J\left(\xi^{(n)}\right) d^{(n)}d\Gamma, \\[2mm] \displaystyle\int_{\Gamma_2} \nabla J\left(\xi^{(n)} + \gamma^{(n)}d^{(n)}\right) d^{(n)}d\Gamma \geq c_2\displaystyle\int_{\Gamma_2} \nabla J\left(\xi^{(n)}\right) d^{(n)}d\Gamma, \end{cases} \qquad (4.17)$$

where ∇J is the gradient of the cost functional; $\xi^{(n)}$ is the n-iteration of the admissible function ξ; and $0 < c_1 < c_2 < 1$. In the case of the conjugate-gradient method, the parameters c_1 and c_2 equal to 0.001 and 0.01, respectively ($c_1 = 0.01$ and $c_2 = 0.9$ in the case of the L-BFGS method; see Sect. 4.7). A search for the descent step length involves iterative solving the direct and adjoint (see below) problems to determine ∇J (e.g., Fletscher 2000).

The gradient of the cost functional

$$\nabla J\left(\xi\right) = \left. \left(k\left(T_\xi\right)\,\frac{\partial z}{\partial \mathbf{n}}\right)\right|_{\Gamma_2} \qquad (4.18)$$

can be found as the solution to the adjoint problem

$$\nabla \cdot \left(\eta\,\left(T_\xi\right)\left(\nabla \mathbf{w} + \nabla \mathbf{w}^T\right)\right) = \nabla q + z\nabla T_\xi, \qquad (4.19)$$

$$\nabla \cdot \mathbf{w} = 0, \qquad (4.20)$$

$$\nabla \cdot \left(\kappa\,\left(T_\xi\right)\nabla z\right) + \left\langle \mathbf{u}_\xi, \nabla z\right\rangle + Ra\,\left\langle \mathbf{e}_2, \mathbf{w}\right\rangle$$

$$= \kappa'\left(T_\xi\right)\left\langle \nabla T_\xi, \nabla z\right\rangle + \eta'\left(T_\xi\right)\left[\left(\nabla \mathbf{w} + \nabla \mathbf{w}^T\right), \nabla \mathbf{u}_\xi\right], \qquad (4.21)$$

with the following boundary conditions

$$\Gamma_1:\quad z = 0,\ \ \mathbf{w} = 0, \qquad (4.22)$$

$$\Gamma_2:\quad z = 0,\ \ \mathbf{w} = 0, \qquad (4.23)$$

$$\Gamma_3:\quad z = 0,\ \ \tilde{\sigma}\mathbf{n} = 0,\ \ q = 0, \qquad (4.24)$$

$$\Gamma_4:\quad z = 2\left(k\left(T_\xi\right)\frac{\partial T_\xi}{\partial \mathbf{n}} - \varphi\right),\quad \left\langle \mathbf{w}, \mathbf{n}\right\rangle = 0,\ \ \tilde{\sigma}\mathbf{n} - \left\langle \tilde{\sigma}\,\mathbf{n}, \mathbf{n}\right\rangle \mathbf{n} = 0, \qquad (4.25)$$

where $\tilde{\sigma} = \eta\left(\nabla \mathbf{w} + \nabla \mathbf{w}^T\right)$; the square brackets $[\mathbf{A}, \mathbf{B}] = \displaystyle\sum_{i,j=1}^{m} a_{ij}\,b_{ij}$ denote the convolution of two $m \times m$ matrices $\mathbf{A} = \left(a_{ij}\right)$ and $\mathbf{B} = \left(b_{ij}\right)$; and sign $'$ means the derivation. The solution is a triplet $(z, \mathbf{w}, q)$ of quasi-temperature (z), quasi-velocity $(\mathbf{w})$, and quasi-pressure q. The derivation of the adjoint problem (4.19), (4.20), (4.21), (4.22), (4.23), (4.24) and (4.25) is presented in Sect. 4.5.

The algorithm for solving the principal problem can be presented using the following steps (the guess function $\xi^{(1)} = \xi^{(1)}(\mathbf{x}) \in \Xi$ determined at Γ_2 is prescribed at the initial iteration):

- *Step 1.* Consider $\xi^{(i)} = \xi^{(i)}(\mathbf{x}), \mathbf{x} \in \Gamma_2$ $(i = 1, 2, \ldots)$ as the boundary condition (4.9) of the auxiliary problem (Eqs. 4.1, 4.2, 4.3 and 4.8, 4.9, 4.10, 4.11) and determine the solution $\left(T_{\xi^{(i)}}, \mathbf{u}_{\xi^{(i)}}, p_{\xi^{(i)}}\right)$ of this problem in Ω.
- *Step 2.* Insert the components $T_{\xi^{(i)}}$ and $\mathbf{u}_{\xi^{(i)}}$ of the solution into the adjoint problem (Eqs. 4.19, 4.20, 4.21, 4.22, 4.23, 4.24, and 4.25) and determine the solution $\left(z = z_{\xi^{(i)}}, \mathbf{w} = \mathbf{w}_{\xi^{(i)}}, q = q_{\xi^{(i)}}\right)$ of this adjoint problem in Ω.
- *Step 3.* Determine the gradient of the cost functional $\nabla J\left(\xi^{(i)}\right)$ from Eq. (4.18) as well as $d^{(i)}$, $\beta^{(i)}$, and $\gamma^{(i)}$ from the conditions (4.15), (4.16), and (4.17), respectively.
- *Step 4.* Determine the value $\xi^{(i+1)}$ from Eq. (4.14).
- *Step 5.* If $J\left(\xi^{(i+1)}\right) + \left\|\nabla J\left(\xi^{(i+1)}\right)\right\|^2 < \varepsilon$, where $\varepsilon > 0$ is a given small number, terminate the minimization problem. Otherwise, the procedure is repeated until the inequality is satisfied.

The performance of the algorithm is evaluated in terms of the number of iterations n required to achieve a prescribed relative reduction of $\xi^{(n)}$. Figure 4.2 presents the evolution of the cost functional $J(\xi^{(n)})$ and the norm of the gradient of the objective functional $\left\|\nabla J\left(\xi^{(n)}\right)\right\| = \left(\int_{\Gamma_2} \left(\nabla J\left(\xi^{(n)}\right)\right)^2 d\Gamma\right)^{1/2}$ versus the number of iterations.

Implementation of the minimization algorithm requires the evaluation of both the cost functional (4.12) and its gradient (4.18). Each evaluation of the objective functional requires an integration of the model Eqs. (4.1), (4.2) and (4.3) with the appropriate boundary conditions (4.8), (4.9), (4.10) and (4.11), whereas the gradient is obtained through the integration of the adjoint problem (Eqs. 4.19, 4.20, 4.21, 4.22, 4.23, 4.24 and 4.25). Thus, the solution of the minimization problem is reduced to solutions of series of well-posed (direct and adjoint) problems.

4.5 Adjoint Problem

Here we present the derivation of the adjoint problem. Let the triplet $\left(T_{\xi+\chi}, \mathbf{u}_{\xi+\chi}, p_{\xi+\chi}\right)$ be the solution of the auxiliary problem (4.1), (4.2), (4.3), (4.8), (4.9), (4.10) and (4.11) for the prescribed condition $T = T_2 = \xi + \chi$ at the boundary Γ_2 (see Eq. 4.9) and the triplet $(T_\xi, \mathbf{u}_\xi, p_\xi)$ be the solution of the same problem for the prescribed condition $T = T_2 = \xi$ at the same boundary, where χ

Fig. 4.2 Reductions of the objective functional (*dashed line 1*) and the norm of the gradient of the objective functional (*solid line 2*) as functions of the number of iterations

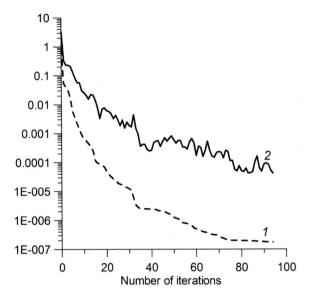

is an admissible increment of the boundary element ξ. The difference of the two solutions $T = T_{\xi+\chi} - T_\xi$, $\mathbf{u} = \mathbf{u}_{\xi+\chi} - \mathbf{u}_\xi$, and $p = p_{\xi+\chi} - p_\xi$ should satisfy the following boundary value problem for $\mathbf{x} \in \Omega$:

$$\nabla \cdot \left(\delta\eta\left(T_\xi\right)\left(\nabla\mathbf{u} + \nabla\mathbf{u}^T\right)\right) + \nabla \cdot \left(\eta\left(T_\xi\right)\left(\nabla\mathbf{u} + \nabla\mathbf{u}^T\right)\right)$$

$$+\nabla \cdot \left(\delta\eta\left(T_\xi\right)\left(\nabla\mathbf{u}_\xi + \nabla\mathbf{u}_\xi^T\right)\right) = \nabla p - Ra\, T\, \mathbf{e}_2, \tag{4.26}$$

$$\nabla \cdot \mathbf{u} = 0, \tag{4.27}$$

$$\nabla \cdot \left(\delta\kappa\left(T_\xi\right)\nabla T\right) + \nabla \cdot \left(\kappa\left(T_\xi\right)\nabla T\right) + \nabla \cdot \left(\delta\kappa\left(T_\xi\right)\nabla T_\xi\right)$$

$$= \langle\mathbf{u}, \nabla T\rangle + \langle\mathbf{u}_\xi, \nabla T\rangle + \langle\mathbf{u}, \nabla T_\xi\rangle, \tag{4.28}$$

with the following boundary conditions

$$\Gamma_1: \quad T = 0, \ \mathbf{u} = 0, \tag{4.29}$$

$$\Gamma_2: \quad T = \chi, \ \mathbf{u} = 0, \tag{4.30}$$

$$\Gamma_3: \quad T = 0, \ \sigma\,\mathbf{n} = 0, \ p = 0, \tag{4.31}$$

$$\Gamma_4: \quad T = 0, \quad \langle \mathbf{u}, \mathbf{n} \rangle = 0, \quad \sigma\mathbf{n} - \langle \sigma\,\mathbf{n}, \mathbf{n} \rangle\,\mathbf{n} = 0, \tag{4.32}$$

where $\delta\eta\left(T_\xi\right) = \eta\left(T_{\xi+\chi}\right) - \eta\left(T_\xi\right)$ and $\delta\kappa\left(T_\xi\right) = \kappa\left(T_{\xi+\chi}\right) - \kappa\left(T_\xi\right)$. We note that

$$J\left(\xi + \chi\right) - J\left(\xi\right) = \int_{\Gamma_4} \left(k\left(T_{\xi+\chi}\right)\frac{\partial T_{\xi+\chi}}{\partial \mathbf{n}} - \varphi\right)^2 d\Gamma - \int_{\Gamma_4} \left(k\left(T_\xi\right)\frac{\partial T_\xi}{\partial \mathbf{n}} - \varphi\right)^2 d\Gamma$$

$$= 2\int_{\Gamma_4} \left(k\left(T_{\xi+\chi}\right)\frac{\partial T_{\xi+\chi}}{\partial \mathbf{n}} - k\left(T_\xi\right)\frac{\partial T_\xi}{\partial \mathbf{n}}\right)\left(k\left(T_\xi\right)\frac{\partial T_\xi}{\partial \mathbf{n}} - \varphi\right)d\Gamma$$

$$+ \int_{\Gamma_4}\left(k\left(T_{\xi+\chi}\right)\frac{\partial T_{\xi+\chi}}{\partial \mathbf{n}} - k\left(T_\xi\right)\frac{\partial T_\xi}{\partial \mathbf{n}}\right)^2 d\Gamma = 2\int_{\Gamma_4}\left(k\left(T_{\xi+\chi}\right)\frac{\partial T_{\xi+\chi}}{\partial \mathbf{n}} - k\left(T_\xi\right)\frac{\partial T_\xi}{\partial \mathbf{n}}\right)$$

$$\times\left(k\left(T_\xi\right)\frac{\partial T_\xi}{\partial \mathbf{n}} - \varphi\right)d\Gamma + o\left(\|\chi\|\right),$$

and accounting for $k\left(T_{\xi+\chi}\right) = k\left(T_\xi\right) + k'\left(T_\xi\right)\ T + o\left(\|T\|\right) = k\left(T_\xi\right) + k'\left(T_\xi\right)\ T + o\left(\|\chi\|\right)$, we obtain

$$2\int_{\Gamma_4}\left(k\left(T_{\xi+\chi}\right)\frac{\partial T_{\xi+\chi}}{\partial \mathbf{n}} - k\left(T_\xi\right)\frac{\partial T_\xi}{\partial \mathbf{n}}\right)\left(k\left(T_\xi\right)\frac{\partial T_\xi}{\partial \mathbf{n}} - \varphi\right)d\Gamma + o\left(\|\chi\|\right)$$

$$= 2\int_{\Gamma_4}\left(k\left(T_\xi\right)\frac{\partial T}{\partial \mathbf{n}} + k'\left(T_\xi\right)\ T\ \frac{\partial T_\xi}{\partial \mathbf{n}} + k'\left(T_\xi\right)\ T\ \frac{\partial T}{\partial \mathbf{n}} + o\left(\|\chi\|\right)\right)$$

$$\times\left(k\left(T_\xi\right)\frac{\partial T_\xi}{\partial \mathbf{n}} - \varphi\right)d\Gamma + o\left(\|\chi\|\right)$$

$$= 2\int_{\Gamma_4}\left(k\left(T_\xi\right)\frac{\partial T}{\partial \mathbf{n}} + k'\left(T_\xi\right)T\frac{\partial T_\xi}{\partial \mathbf{n}} + o\left(\|\chi\|\right)\right)\left(k\left(T_\xi\right)\frac{\partial T_\xi}{\partial \mathbf{n}} - \varphi\right)d\Gamma + o\left(\|\chi\|\right)$$

$$= 2\int_{\Gamma_4}\left(k\left(T_\xi\right)\frac{\partial T}{\partial \mathbf{n}} + k'\left(T_\xi\right)\ T\ \frac{\partial T_\xi}{\partial \mathbf{n}}\right)\left(k\left(T_\xi\right)\frac{\partial T_\xi}{\partial \mathbf{n}} - \varphi\right)d\Gamma + o\left(\|\chi\|\right),$$

and hence

$$J\left(\xi + \chi\right) - J\left(\xi\right)$$
$$= \int_{\Gamma_4}\left(k\left(T_\xi\right)\frac{\partial T}{\partial \mathbf{n}} + k'\left(T_\xi\right)\ T\ \frac{\partial T_\xi}{\partial \mathbf{n}}\right)2\left(k\left(T_\xi\right)\frac{\partial T_\xi}{\partial \mathbf{n}} - \varphi\right)d\Gamma + o\left(\|\chi\|\right).$$

$$\tag{4.33}$$

A test function $\mathbf{w} = \mathbf{w}(\mathbf{x})$, $\mathbf{x} \in \Omega$, is assumed to satisfy the incompressibility condition

$$\nabla \cdot \mathbf{w} = 0, \tag{4.34}$$

and the boundary conditions

$$\Gamma_1 \text{ and } \Gamma_2 : \quad \mathbf{w} = 0, \tag{4.35}$$

$$\Gamma_3 : \quad \tilde{\sigma}\, \mathbf{n} = 0, \tag{4.36}$$

$$\Gamma_4 : \quad \langle \mathbf{w}, \mathbf{n} \rangle = 0, \quad \tilde{\sigma}\mathbf{n} - \langle \tilde{\sigma}\,\mathbf{n}, \mathbf{n} \rangle \mathbf{n} = 0. \tag{4.37}$$

Multiplying Eq. (4.26) by a test function $\mathbf{w} = \mathbf{w}(\mathbf{x})$, integrating the resultant equation over Ω, considering Eqs. (4.34), (4.35), (4.36) and (4.37) and after integration by parts, the following equation is obtained:

$$\int_{\Omega} \langle \mathbf{u}, \nabla \cdot (\eta\,(T_{\xi})\,(\nabla\mathbf{w} + \nabla\mathbf{w}^T)) \rangle\, dx - \int_{\Omega} \eta'\,(T_{\xi})\; T\,\left[\nabla\mathbf{w} + \nabla\mathbf{w}^T, \nabla\mathbf{u}_{\xi}\right]\, dx$$

$$+ \int_{\Omega} Ra\; T\; \langle \mathbf{w}, \mathbf{e}_2 \rangle\, dx = o\,(\,\|\chi\|\,), \tag{4.38}$$

where the relation $\left[\nabla\mathbf{w} + \nabla\mathbf{w}^T, \nabla\mathbf{u}_{\xi}\right]$ can be represented in a symmetric form as $\left[\nabla\mathbf{w} + \nabla\mathbf{w}^T, \nabla\mathbf{u}_{\xi} + \nabla\mathbf{u}_{\xi}^T\right]/2$. Multiply Eq. (4.27) by a test scalar function $q = q(\mathbf{x})$, $\mathbf{x} \in \Omega$, and integrate by parts the resultant equation over Ω. Assuming that the function $q = 0$ at Γ_3 and considering boundary conditions (4.29), (4.30), (4.31) and (4.32) for the vector function $\mathbf{u}$:

$$\int_{\Omega} \langle \mathbf{u}, \nabla q \rangle\, dx = 0. \tag{4.39}$$

Multiply Eq. (4.28) by a test scalar function $z = z(\mathbf{x})$, $\mathbf{x} \in \Omega$, and integrate by parts the resultant equation over Ω. Considering boundary conditions (4.29), (4.30), (4.31) and (4.32) for the function T and assuming that the function z satisfies the following boundary conditions: $z = 0$ at Γ_1, Γ_2, and Γ_3, and $z = 2\left(k\,(T_{\xi})\,\frac{\partial T_{\xi}}{\partial \mathbf{n}} - \varphi\right)$ at Γ_4, the modified equation can be presented as

$$\int_{\Omega} T \left\{ \nabla \cdot \left(\kappa \left(T_{\xi} \right) \nabla z \right) - \kappa' \left(T_{\xi} \right) \left\langle \nabla T_{\xi}, \nabla z \right\rangle + \left\langle \mathbf{u}_{\xi}, \nabla z \right\rangle \right\} \, dx - \int_{\Omega} \left\langle \mathbf{u}, \nabla T_{\xi} \right\rangle z \, dx$$

$$+ \int_{\Gamma_4} \left(\kappa \left(T_{\xi} \right) \frac{\partial T}{\partial \mathbf{n}} + \kappa' \left(T_{\xi} \right) T \frac{\partial T_{\xi}}{\partial \mathbf{n}} \right) z \, d\Gamma - \int_{\Gamma_2} \kappa \left(T_{\xi} \right) \frac{\partial z}{\partial \mathbf{n}} \chi \, d\Gamma = o \left(\| \chi \| \right).$$

$$(4.40)$$

Now add Eq. (4.40) to Eq. (4.39) and then deduct Eq. (4.40):

$$\int_{\Omega} \left\langle \mathbf{u}, \left\{ \nabla \cdot \left(\eta \left(T_{\xi} \right) \left(\nabla \mathbf{w} + \nabla \mathbf{w}^T \right) \right) - z \nabla T_{\xi} - \nabla q \right\} \right\rangle dx$$

$$+ \int_{\Omega} T \left\{ \nabla \cdot \left(\kappa \left(T_{\xi} \right) \nabla z \right) - \kappa' \left(T_{\xi} \right) \left\langle \nabla T_{\xi}, \nabla z \right\rangle + \left\langle \mathbf{u}_{\xi}, \nabla z \right\rangle \right.$$

$$\left. - \eta' \left(T_{\xi} \right) \left[\nabla \mathbf{w} + \nabla \mathbf{w}^T, \nabla \mathbf{u}_{\xi} \right] + Ra \left\langle \mathbf{w}, \mathbf{e}_2 \right\rangle \right\} \, dx$$

$$+ \int_{\Gamma_4} \left(\kappa \left(T_{\xi} \right) \frac{\partial T}{\partial \mathbf{n}} + \kappa' \left(T_{\xi} \right) T \frac{\partial T_{\xi}}{\partial \mathbf{n}} \right) z \, d\Gamma - \int_{\Gamma_2} \kappa \left(T_{\xi} \right) \frac{\partial z}{\partial \mathbf{n}} \chi \, d\Gamma = o \left(\| \chi \| \right).$$

$$(4.41)$$

Assuming that the expression in braces in Eq. (4.41) equals to zero, Eqs. (4.19) and (4.21) and the equality for two boundary integrals:

$$\int_{\Gamma_4} \left(k \left(T_{\xi} \right) \frac{\partial T}{\partial \mathbf{n}} + k' \left(T_{\xi} \right) T \frac{\partial T_{\xi}}{\partial \mathbf{n}} \right) z \, d\Gamma = \int_{\Gamma_2} k \left(T_{\xi} \right) \frac{\partial z}{\partial \mathbf{n}} \chi \, d\Gamma + o \left(\| \chi \| \right). \quad (4.42)$$

Insert Eq. (4.42) into Eq. (4.34) to derive:

$$J \left(\xi + \chi \right) - J \left(\xi \right) = \int_{\Gamma_2} \nabla J \left(\xi \right) \chi \, d\Gamma + o \left(\| \chi \| \right), \quad \nabla J \left(\xi \right) = k \left(T_{\xi} \right) \left. \frac{\partial z}{\partial \mathbf{n}} \right|_{\Gamma_2}.$$

4.6   Numerical Approach

To implement the algorithm for solving the minimization problem, Korotkii et al. (2016) developed a numerical code using OpenFOAM (http://www.openfoam.org). The mathematical problem was discretized by the finite volume method (e.g. Ismail-Zadeh and Tackley 2010). The model domain was discretized by 1500 hexahedral finite volumes. The SIMPLE method (Patankar and Spalding 1972) for used to determine velocity and pressure at a given temperature. To implement the conjugate-gradient method, a set of linear algebraic equations (SLAEs) with positive-definite and symmetric matrices was solved. In the case of the heat equation, SLAEs were

solved by the biconjugate gradient stabilized method (Van der Vorst 1992) with
the pre-conditioner of incomplete *LU*-decomposition. The linear Gaussian scheme
with a flow control was used to discretize the Laplace operator. To approximate the
convective operator, we employed the total variation diminishing (TVD) method,
which gives a second-order accurate solution, with the minmod limiter (Sweby
1984; Wang and Hutter 2001; Ismail-Zadeh et al. 2007). The relaxation parameters
are 0.7 and 0.3 for the velocity and pressure, respectively; and the relative
accuracy of the SLAE solutions are 10^{-3}. Korotkii et al. (2016) performed model
computations using a single CPU Intel Core i5 2.6 GHz with 16 GB memory, OS X
10.10. An average computational time for 80 iterations in the inverse problem was
75 min: this included the time required for minimization of the cost functional by
the conjugate gradient method, and the time to solve the direct and adjoint problems
(normally 4–5 iterations) to determine the descent step length.

4.7   Model Results and Discussion

Following Korotkii et al. (2016), we consider here a model of lava advancing down
the slope (Fig. 4.1) and assume that temperature and heat flow are available from
remote thermal measurements. The boundary of the model domain consists of the
following parts: Γ_1 is a line segment connecting points $\mathbf{x}^A = \left(x_1^A, x_2^A\right) = (0, 2.5)$ and
$\mathbf{x}^B = \left(x_1^B, x_2^B\right) = (0, 1.5)$; Γ_2 is a circular arc connecting points $\mathbf{x}^B, \mathbf{x}^C = \left(x_1^C, x_2^C\right) = (1.5, 0.5)$, and $\mathbf{x}^D = \left(x_1^D, x_2^D\right) = (3.0, 0.0)$; Γ_3 is a line segment connecting points
$\mathbf{x}^D$ and $\mathbf{x}^E = \left(x_1^E, x_2^E\right) = (3.0, 0.5)$; and Γ_4 is a circular arc connecting points $\mathbf{x}^E$,
$\mathbf{x}^F = \left(x_1^F, x_2^F\right) = (1.5, 1.2)$ and $\mathbf{x}^A$. The following dimensional parameters are used
in the modelling: $\alpha = 10^{-5}$ K^{-1}, $g = 9.8$ m s^{-2}, $h = 10$ m, $\rho_{ref} = 3000$ kg m^{-3},
$\eta_{ref} = 3.5 \times 10^9$ Pa s, $T_{ref} = 300$ K, $T_* = 1473$ K, $\Delta T = T_* - T_{ref}$, $\kappa_{ref} = 10^{-6}$ m^2
s^{-1}, $c_p = 1200$ J kg^{-1} K^{-1}, and therefore, the Rayleigh number is $Ra = 100$. The
temperature-dependent viscosity (Griffiths 2000) and conductivity (Hidaka et al.
2005) used in this case study are presented as:
$$\eta(T) = \exp\left(n\left(T_* - T_{ref}T\right)\right), n = 1.3 \times 10^{-4} \text{ K}^{-1}, \text{ and}$$

$$k(T) = \begin{cases} 1.15 + 5.9 \cdot 10^{-7}\left(T_{ref}T - T_*\right)^2, T_{ref}T < T_*, \\ 1.15 + 9.7 \cdot 10^{-6}\left(T_{ref}T - T_*\right)^2, T_{ref}T > T_*. \end{cases}$$

At Γ_1 the temperature $T_1(x_1, x_2) = 5.0 - 0.5\left(x_2 - x_2^B\right)$, $x_2 \in \left[x_2^B, x_2^A\right]$, and the
velocity $\mathbf{u}_1(x_2) = U(x_2)\mathbf{n}_1$ are prescribed, where $\mathbf{n}_1 = \left(\sqrt{2}/2, -\sqrt{2}/2\right)$ and
$U(x_2)$ is the parabola passing through the following three points: $U\left(x_2^A\right) = 10$,
$U\left(x_2^B\right) = 0$, and $U\left(0.5\left(x_2^A + x_2^B\right)\right) = 7.25$. The temperature is $T_3(x_1, x_2) = 3.5 - 2\left(x_2 - x_2^D\right)$, $x_2 \in \left[x_2^D, x_2^E\right]$ and $T_4(x_1, x_2) = 4.5 - 2\left(x_1 - x_1^A\right)/3$, $x_1 \in \left[x_1^A, x_1^E\right]$ at
Γ_3 and Γ_4, respectively. Considering guess temperature $\xi^{(1)} = \xi^{(1)}(\mathbf{x})$ at Γ_2, the

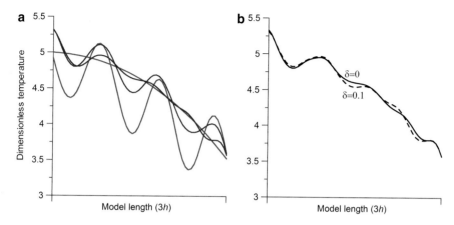

Fig. 4.3 Reconstruction of the temperature at the boundary Γ_2 (**a**). The *red curve* corresponds to the target temperature, the *green curve* to the guess temperature, the *brown curve* to the temperature after 5 iterations, and the *blue curve* to the temperature after the 10 iterations. The reconstructed temperature after 10 iterations (**b**) in the case of no noise in the heat flow at the upper boundary of the lava (*solid line*; the *blue curve* in **a**) and in the case of the noise magnitude $\delta = 0.1$ in the heat flow (*dashed line*)

algorithm described in Sect. 4.4 is used to find the temperature at Γ_2 and hence solve the problem (4.1), (4.2), (4.3), (4.4), (4.5) and (4.7).

The cost functional is reduced to about 10^{-5} after 30 iterations (Fig. 4.2). The reconstruction of the temperature at the boundary Γ_2 versus the number of iterations is presented in Fig. 4.3a. The number of iterations to get a given accuracy in reduction of the cost functional depends on the initial 'guess' temperature at Γ_2. The closer is the guess temperature to the target temperature, the less number of iterations is needed.

Figure 4.4 shows the reconstruction process of the lava temperature and flow velocity from the initial iteration to the 80th iteration. The temperature and velocity residuals, that is, the difference between the temperature and velocities predicted by the forward model (with the target temperature at Γ_2) and those reconstructed, are also presented in Fig. 4.4. The results of this modelling show that the restoration works quite well: the temperature residuals are very low already after 80 iterations within the almost entire model domain.

The accuracy of temperature measurements and inferred heat flux density can be attributed to the accuracy of the calibration curve of remote sensors and the noise of the sensors. Considering these sources of errors of measured temperatures, the errors would range from 0.1 to 1 K (Short and Stuart 1983). The heat flow errors inferred from the Stefan-Bolzmann law can be then estimated between 0.6 and 6 W m^{-2} at the reference temperature $T_{ref} = 300$ K, which are related to dimensionless error values from 0.0013 to 0.013 (normalized with respect to heat flow at the reference temperature).

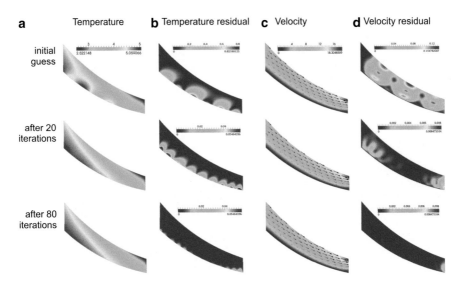

Fig. 4.4 Reconstruction of the lava temperature (**a**) and the flow velocity (**c**) after 20 and 80 iterations. The relevant residuals of the temperature (**b**) and the velocity (**d**) indicate the quality of the reconstruction

Several numerical experiments have been performed by Korotkii et al. (2016) where a noise on the 'measured' data was introduced. Particularly, a disturbance on the heat flow $\varphi(\cdot)$ at the boundary Γ_4 is presented as $\varphi_\delta(\cdot) = \varphi(\cdot) + \delta\gamma(\cdot)$, where δ is the magnitude of the disturbance; $\gamma(\cdot)$ is the function generating numbers that are uniformly distributed over the interval $[-1, 1]$; and $\varphi(\cdot)$ is obtained from the solution of Eqs. (4.1), (4.2) and (4.3) with the conditions at the boundaries (4.8), (4.9), (4.10) and (4.11) for $T_2 = \xi^{(1)}$ at Γ_2. Three values for the noise magnitude δ (0.001, 0.01, and 0.1) are chosen to approximate the possible noise level of the remote thermal measurements.

The influence of the noise on the reconstruction of the temperature and flow velocity has been analysed by Korotkii et al. (2016). The computations show that the errors (temperature and velocity residuals, Fig. 4.5) get larger with increase of the noise of the input data. Meanwhile for some range of the noise ($\delta \leq 0.01$) the reconstructions are still reasonable as the temperature and velocity residuals are not high (Fig. 4.5). Namely, if $M_T = \max_{\mathbf{x}\in\Omega} \left| T_{30}(\mathbf{x}) - T^0(\mathbf{x}) \right|$ and $M_{\mathbf{u}} = \max_{\mathbf{x}\in\Omega} \left\| \mathbf{u}_{30}(\mathbf{x}) - \mathbf{u}^0(\mathbf{x}) \right\|_{\mathbb{R}^2}$, where $T^0(\mathbf{x})$ and $\mathbf{u}^0(\mathbf{x})$ are the solution of the direct problem (4.1), (4.2) and (4.3) and (4.8), (4.9), (4.10) and (4.11), than $M_T = 0.095$, 0.096, 0.099, and 0.265 and $M_{\mathbf{u}} = 0.0073$, 0.0074, 0.0075, and 0.01526 for $\delta = 0$, 0.001, 0.01, and 0.1, respectively. To test a sensitivity of the approach to changes in flow patterns, the magnitude of the velocity $|U|$ has been varied at the left-side boundary Γ_1 of the model domain between 1 and 25, and the Rayleigh number between 1 and 10,000. The approach is rather robust to changes in the velocity magnitude and in Ra.

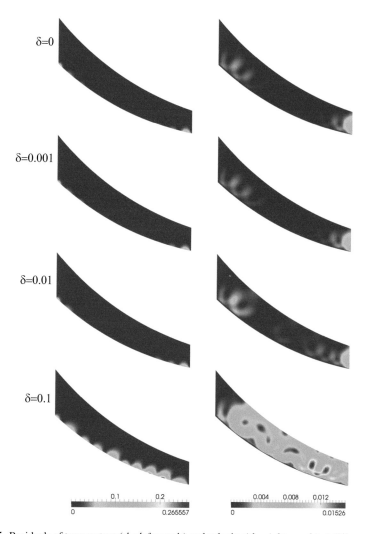

Fig. 4.5 Residuals of temperature (*the left panels*) and velocity (*the right panels*) at different noise magnitudes δ

A performance of the numerical approach depends on optimization methods. In the presented approach the conjugate-gradient method was used. To compare its performance with other optimization methods, the limited-memory BFGS method (or L-BFGS method) was employed. This method belongs to the family of quasi-Newton methods (e.g., Nocedal and Wright 1999). To minimize the cost functional (4.12) using the L-BFGS method, components $d^{(n)}$ in Eq. (4.14) are determined as $d^{(1)} = -\nabla J\left(\xi^{(1)}\right)$ and $d^{(n)} = -\mathbf{B}^{(n)}\nabla J\left(\xi^{(n)}\right)$ ($n = 2,\ 3,\ \ldots$), where $\mathbf{B}^{(n)}$ is the approximated inverse Hessian operator. When the L-BFGS method is used, the average computational time to perform 80 iterations for minimization of the

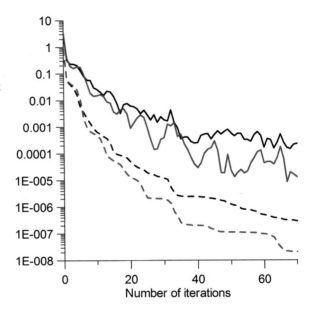

Fig. 4.6 Reductions of the objective functional (*dashed lines*) and the norm of the gradient of the objective functional (*solid lines*) in the case of the conjugate gradient method (*black lines*) and in the case of the L-BFGS method (*red lines*)

cost functional is reduced to 15 min (by the factor of 5) compared to the case of the conjugate-gradient method used. The computational time reduction is achieved because the descent step size in the iteration scheme is determined much faster. The reduction of the objective functional and the norm of the gradient of the objective functional with the number of iterations is faster than in the case of the conjugate-gradient method (Fig. 4.6).

Rather accurate reconstruction of the model temperature and flow velocity relies on the chosen methods for minimization of the cost functional (4.12), i.e. the Polak-Ribière conjugate-gradient method or the L-BFGS method. In the general case, a Tikhonov regularization term should be introduced in the cost functional as:

$$J_\alpha\left(\xi\right) = \int_{\Gamma_4}\left(k\left(T_\xi\right)\frac{\partial T_\xi}{\partial \mathbf{n}} - \varphi_\delta\right)^2 d\Gamma + \alpha\int_{\Gamma_2}\xi^2\, d\Gamma.$$

Here $\alpha > 0$ is a small regularization parameter, and φ_δ is the measured heat flow with a measurement error δ. The regularization term should account for *a priori* information on the problem's solution (e.g., its monotony property, maximum and minimum values, and the total variation diminishing). The introduction of the regularization term in the cost functional makes the minimization problem more stable and less dependent on measurement errors. For a suitable regularization parameter $\alpha = \alpha\left(\delta\right)$, the minimum of the regularized cost functional will tend to the minimum of the functional (4.12) at $\delta \to 0$ (Tikhonov and Arsenin 1977). In reality the choice of the regularization parameter is a challenging issue as it depends on several factors, e.g., on errors of measured data (e.g., Kabanikhin 2011). Note that the gradient of the regularized cost functional is represented as $\nabla J_\alpha(\xi) = \nabla J_0(\xi) + 2\alpha\xi$, and the Hessian matrix is positive ($\nabla^2 J_\alpha > 0$) at $\alpha > 0$.

There are several simplifications in the presented model of lava flow that can be overcome in future, but require further development of the algorithm and increase in computational resources. For example, the proposed numerical approach allows also for reconstructing the temperature at the right boundary of the model domain (if heat flux is negligible at its lower boundary) or at lower and right boundaries simultaneously. The problem presented here can be extended to the non-steady-state flow, but this will complicate the mathematical and computational approaches. Meanwhile, as the measurements on absolute temperature are discrete in time (e.g., depending on the location of Landsat satellites), a problem of non-stationary flow can be reduced to a number of steady-state flow problems with varying boundary conditions at the upper model surface (where the discrete-in-time measurement are available). A more complicated lava rheology with formation and disintegration of solid crust (e.g., Tsepelev et al. 2016) can be considered. The influence of the shape of the crust and the degree of its disintegration on the radiated heat flux (Neri 1998) can be significant. The application of the VAR method to lava flow and its numerical implementation have a wide range of applications in other problems of reconstruction of the flows of fluids with strongly temperature dependent viscosity, for example, in chemical technology or oil industry.

References

Costa A, Macedonio G (2005a) Computational modeling of lava flows: a review. In: Manga M, Ventura G (eds) Kinematics and dynamics of lava flows. Geological Society of America Special Papers 396, Boulder, pp 209–218

Costa A, Macedonio G (2005b) Numerical simulation of lava flows based on depth-averaged equations. Geophys Res Lett 32:L05304. doi:10.1029/2004GL021817

Cutter S, Ismail-Zadeh A, Alcántara-Ayala I, Altan O, Baker DN, Briceño S, Gupta H, Holloway A, Johnston D, McBean GA, Ogawa Y, Paton D, Porio E, Silbereisen RK, Takeuchi K, Valsecchi GB, Vogel C, Wu G (2015) Pool knowledge to stem losses from disasters. Nature 522:277–279

Fletscher R (2000) Practical methods of optimization, 2nd edn. Wiley, Chichester

Flynn LP, Harris AJL, Wright R (2001) Improved identification of volcanic features using Landsat 7 ETM+. Remote Sens Environ 78:180–193

Griffiths RW (2000) The dynamics of lava flows. Annu Rev Fluid Mech 32:477–518

Harris AJL, Flynn LP, Matias O, Rose WI, Cornejo J (2004) The evolution of an active silicic lava flow field: an ETM+ perspective. J Volcanol Geotherm Res 135:147–168

Harris AJL, Dehn J, Calvari S (2007) Lava effusion rate definition and measurement: a review. Bull Volcanol 70:1–22

Hidaka M, Goto A, Umino S, Fujita E (2005) VTFS project: development of the lava flow simulation code LavaSIM with a model for three-dimensional convection, spreading, and solidification. Geochem Geophys Geosyst 6:Q07008. doi:10.1029/2004GC000869

Ishihara K, Iguchi M, Kamo K (1989) Numerical simulation of lava flows on some volcanoes in Japan. In: Fink J (ed) Lava flows and domes, vol 2, IAVCEI Proc. Volcanol. Springer, New York

Ismail-Zadeh A, Tackley P (2010) Computational methods for geodynamics. Cambridge University Press, Cambridge

Ismail-Zadeh A, Korotkii A, Schubert G, Tsepelev I (2007) Quasi-reversibility method for data assimilation in models of mantle dynamics. Geophys J Int 170:1381–1398

Kabanikhin SI (2011) Inverse and ill-posed problems. Theory and applications. De Gruyter, Berlin

Korotkii AI, Kovtunov DA (2006) Reconstruction of boundary regimes in an inverse problem of thermal convection of a high viscous fluid. Proc Inst Math Mech Ural Branch Russ Acad Sci 12(2):88–97 (in Russian)

Korotkii AI, Starodubtseva YV (2014) Direct and inverse problems for models of stationary reactive-convective-diffusive flow. Proc Inst Math Mech Ural Branch Russ Acad Sci 20(3):98–113 (in Russian)

Korotkii A, Kovtunov D, Ismail-Zadeh A, Tsepelev I, Melnik O (2016) Quantitative reconstruction of thermal and dynamic characteristics of lava flow from surface thermal measurements. Geophys J Int. doi:10.1093/gji/ggw117

Ladyzhenskaya OA (1969) The mathematical theory of viscous incompressible flow. Gordon and Breach, New York

Lions JL (1971) Optimal control of systems governed by partial differential equations. Springer, Berlin/Heidelberg

Miyamoto H, Sasaki S (1998) Numerical simulations of flood basalt lava flows: roles of parameters on lava flow morphologies. J Geophys Res 103:27489–27502

Neri A (1998) A local heat transfer analysis of lava cooling in the atmosphere: application to thermal diffusion-dominated lava flows. J Volcanol Geotherm Res 81:215–243

Nocedal J, Wright SJ (1999) Numerical optimization. Springer, New York

Patankar SV, Spalding DB (1972) A calculation procedure for heat and mass transfer in three-dimensional parabolic flows. Int J Heat Mass Transf 15:1787–1806

Polak E (1997) Optimization: algorithms and consistent approximations. Springer, Berlin/Heidelberg

Short NM, Stuart LM (1983) The heat capacity mapping mission (HCMM) anthology. Scientific and Technical Information Branch, National Aeronautics & Space Administration, Washington, DC

Sweby PK (1984) High resolution schemes using flux limiters for hyperbolic conservation laws. J Numer Anal 21:995–1011

Temam R (1977) Navier-Stokes equations: theory and numerical analysis. North-Holland, Amsterdam

Tikhonov AN, Arsenin VY (1977) Solution of ill-posed problems. Winston, Washington, DC

Tsepelev I, Ismail-Zadeh A, Melnik O, Korotkii A (2016) Numerical modelling of fluid flow with rafts: an application to lava flows. J Geodyn. doi:10.1016/j.jog.2016.02.010

Van der Vorst HA (1992) BI-CGSTAB: a fast and smoothly converging variant of BI-CG for the solution of nonsymmetric linear systems. SIAM J Sci Stat Comput 13(2):631–644

Wang Y, Hutter K (2001) Comparison of numerical methods with respect to convectively dominated problems. Int J Numer Methods Fluids 37:721–745

Chapter 5
Quasi-Reversibility Method and Its Applications

Abstract In this chapter, we introduce a quasi-reversibility (QRV) approach to data assimilation, which allows for incorporating observations (at present) and unknown initial conditions (in the past) for physical parameters (e.g., temperature and flow velocity) into a three-dimensional dynamic model in order to determine the initial conditions. The dynamic model is described by the backward heat, motion, and continuity equations. The use of the QRV method implies the introduction into the backward heat equation of the additional term involving the product of a small regularization parameter and a higher order temperature derivative. The data assimilation in this case is based on a search of the best fit between the forecast model state and the observations by minimizing the regularization parameter. We present the application of the QRV method to two case studies: evolution of (i) mantle plumes and (ii) a relic lithospheric slab.

Keywords Quasi-reversibility • Regularization • Mantle plume • Lithosphere • Slab sinking • Vrancea • Numerical modelling

5.1 Basic Idea of the Quasi-Reversibility (QRV) Method

The principal idea of the QRV method is based on the transformation of an ill-posed problem into a well-posed problem (Lattes and Lions 1969). In the case of the backward heat equation, this implies an introduction of an additional term into the equation, which involves the product of a small regularization parameter and higher order temperature derivative. The additional term should be sufficiently small compared to other terms of the heat equation and allow for simple additional boundary conditions. The data assimilation in this case is based on a search of the best fit between the forecast model state and the observations by minimizing the regularization parameter. The QRV method is proven to be well suited for smooth and non-smooth input data (Lattes and Lions 1969; Samarskii and Vabishchevich 2007).

To explain the transformation of the problem, we follow Ismail-Zadeh et al. (2007) and consider the following boundary-value problem for the one-dimensional heat conduction problem

© The Author(s) 2016
A. Ismail-Zadeh et al., *Data-Driven Numerical Modelling in Geodynamics:
Methods and Applications*, SpringerBriefs in Earth Sciences,
DOI 10.1007/978-3-319-27801-8_5

59

$$\frac{\partial T\,(t,x)}{\partial t} = \frac{\partial^2 T\,(t,x)}{\partial x^2}, \quad 0 \le x \le \pi, \quad 0 \le t \le t^*, \tag{5.1}$$

$$T\,(t, x = 0) = T\,(t, x = \pi) = 0, \quad 0 \le t \le t^*, \tag{5.2}$$

$$T\,(t = 0, x) = \frac{1}{4n + 1}\sin\left((4n + 1)\,x\right), \quad 0 \le x \le \pi. \tag{5.3}$$

The analytical solution to (5.1), (5.2) and (5.3) can be obtained in the following form

$$T\,(t, x) = \frac{1}{4n + 1}\exp\left(-(4n + 1)^2 t\right)\,\sin\left((4n + 1)\,x\right). \tag{5.4}$$

Figure 5.1 presents the solution (solid curves) for time interval $0 \le t \le t^* = 0.14$ and $n = 1$.

It is known that the backward heat conduction problem is ill-posed (e.g. Kirsch 1996). To transform the problem into a well-posed problem, a term is introduced in Eq. (5.1) involving the product of a small parameter $\beta > 0$ and higher order temperature derivative:

$$\frac{\partial T_\beta\,(t,x)}{\partial t} = \frac{\partial^2 T_\beta\,(t,x)}{\partial x^2} - \beta\frac{\partial^4}{\partial x^4}\left(\frac{\partial T_\beta\,(t,x)}{\partial t}\right), \quad 0 \le x \le \pi, \quad 0 \le t \le t^*, \tag{5.5}$$

$$T_\beta\,(t, x = 0) = T_\beta\,(t, x = \pi) = 0, \quad 0 \le t \le t^*, \tag{5.6}$$

$$\frac{\partial^2 T_\beta\,(t, x = 0)}{\partial x^2} = \frac{\partial^2 T_\beta\,(t, x = \pi)}{\partial x^2} = 0, \quad 0 \le t \le t^*, \tag{5.7}$$

$$T_\beta\,(t = t^*, x) = \frac{1}{4n + 1}\exp\left(-(4n + 1)^2 t^*\right)\,\sin\left((4n + 1)\,x\right), \quad 0 \le x \le \pi. \tag{5.8}$$

Here the initial condition is assumed to be the solution (5.4) to the heat conduction problem (5.1), (5.2) and (5.3) at $t = t^*$. The subscript β at T_β is used to emphasize the dependence of the solution to problem (5.5), (5.6), (5.7) and (5.8) on the regularization parameter. The analytical solution to the regularized backward heat conduction problem (5.5), (5.6), (5.7) and (5.8) is represented as:

$$T_\beta\,(t, x) = A_n \exp\left(\frac{-(4n+1)^2 t}{1 + \beta(4n+1)^4}\right)\sin\left((4n + 1)\,x\right),$$

$$A_n = \frac{1}{4n+1}\exp\left(-(4n + 1)^2 t^*\right)\exp^{-1}\left(\frac{-(4n+1)^2 t^*}{1 + \beta(4n+1)^4}\right), \tag{5.9}$$

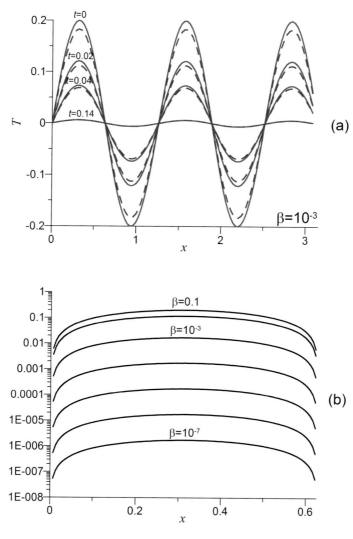

Fig. 5.1 Comparison of the exact solutions to the heat conduction problem (*red curves*; **a** and **b**) and to the regularized backward heat conduction problem (**a**: $\beta = 10^{-3}$ and **b**: $\beta = 10^{-7}$; *blue dashed curves*). The temperature residual between two solutions is presented in panel (**c**) at various values of the regularization parameter β (After Ismail-Zadeh et al. 2007)

and the solution approaches the initial condition for the problem (5.1), (5.2) and (5.3) at $t = 0$ and $\beta \rightarrow 0$. Figure 5.1a illustrates the solution to the regularized problem at $\beta = 10^{-3}$ (dashed curves) and $n = 1$. The temperature residual (Fig. 5.1b) indicates that the solution (5.9) approaches the solution (5.4) with $\beta \rightarrow 0$. Samarskii and Vabischevich (2007) estimated the stability of the solution to problem (5.5), (5.6) and (5.7) with respect to the initial condition expressed in the form $T_\beta\,(t = t^*, x) = T_\beta^*$:

$$\left\| T_\beta \left(t,x \right) \right\| + \beta \left\| \partial T_\beta \left(t,x \right) / \partial x \right\| \leq C \left(\left\| T_\beta^* \right\| + \beta \left\| \partial T_\beta^* / \partial x \right\| \right) \exp \left[\left(t^* - t \right) \beta^{-1/2} \right],$$

where C is a constant, and showed that the natural logarithm of errors will increase in a direct proportion to time and inversely to the root square of the regularization parameter.

Any regularization has its advantages and disadvantages. A regularizing operator is used in a mathematical problem to (i) accelerate a convergence; (ii) fulfil the physical laws (e.g. maximum principal, conversation of energy, etc.) in discrete equations; (iii) suppress a noise in input data and in numerical computations; and (iv) take into account *a priori* information about an unknown solution and hence to improve a quality of computations. The major drawback of regularization is that the accuracy of the solution to a regularized problem is always lower than that to a non-regularized problem.

The transformation to the regularized backward heat problem is not only a mathematical approach to solving ill-posed backward heat problems, but has some physical meaning: it can be explained on the basis of the concept of relaxing heat flux for heat conduction (e.g. Vernotte 1958). The classical Fourier heat conduction theory provides the infinite velocity of heat propagation in a region. The instantaneous heat propagation is unrealistic, because the heat is a result of the vibration of atoms and the vibration propagates in a finite speed (Morse and Feshbach 1953). To accommodate the finite velocity of heat propagation, a modified heat flux model was proposed by Vernotte (1958) and Cattaneo (1958).

The modified Fourier constitutive equation is expressed as $\overrightarrow{Q} = -k\nabla T - \tau \, \partial \overrightarrow{Q} / \partial t$, where $\overrightarrow{Q}$ is the heat flux, and k is the coefficient of thermal conductivity. The thermal relaxation time $\tau = k / \left(\rho c_p v^2 \right)$ is usually recognized to be a small parameter (Yu et al. 2004), where ρ is the density, c_p is the specific heat, and v is the heat propagation velocity. The situation for $\tau \to 0$ leads to instantaneous diffusion at infinite propagation speed, which coincides with the classical thermal diffusion theory. The heat conduction equation $\partial T / \partial t = \nabla^2 T + \tau \, \partial^2 T / \partial t^2$ based on non-Fourier heat flux can be considered as a regularized heat equation. If the Fourier law is modified further by an addition of the second derivative of heat flux, e.g. $\overrightarrow{Q} = -k\nabla T + \beta \frac{\partial^2 \overrightarrow{Q}}{\partial t^2}$, where small β is the relaxation parameter of heat flux (Bubnov 1976, 1981), the heat conduction equation can be transformed into a higher order regularized heat equation similar to Eq. (5.5).

5.2 Mathematical Statement

For convenience, a set of Eqs. (3.1), (3.2) and (3.3) with the relevant boundary and initial conditions is represented as two mathematical problems: the *boundary value problem for the flow velocity* (it includes the Stokes equation, the incompressibility equation subject to appropriate boundary conditions)

$$\nabla P = \nabla \cdot \left[\eta \left(\nabla \mathbf{u} + \nabla \mathbf{u}^T \right) \right] + RaT\mathbf{e}, \qquad \mathbf{x} \in \Omega, \qquad (5.10)$$

$$\nabla \cdot \mathbf{u} = 0, \qquad \mathbf{x} \in \Omega, \qquad (5.11)$$

$$\langle \mathbf{u}, \mathbf{n} \rangle = 0, \quad \sigma \mathbf{n} - \langle \sigma \, \mathbf{n}, \mathbf{n} \rangle \mathbf{n} = 0, \qquad \mathbf{x} \in \partial\Omega, \qquad (5.12)$$

where η is the viscosity, Ra is the Rayleigh number, $\sigma = \eta \left(\nabla \mathbf{u} + \nabla \mathbf{u}^T \right)$ is the stress tensor, and $\mathbf{n}$ is the outward unit normal vector at a point on the boundary $\partial\Omega$; and *the initial-boundary value problem for temperature* (it includes the heat equation subject to appropriate boundary and initial conditions)

$$\partial T / \partial t + \langle \mathbf{u}, \nabla T \rangle = \nabla^2 T, \qquad t \in [0, \vartheta], \qquad \mathbf{x} \in \Omega, \qquad (5.13)$$

$$\sigma_1 T + \sigma_2 \partial T / \partial \mathbf{n} = T_*, \qquad t \in [0, \vartheta], \qquad \mathbf{x} \in \partial\Omega, \qquad (5.14)$$

$$T(0, \mathbf{x}) = T_0(\mathbf{x}), \qquad \mathbf{x} \in \Omega, \qquad (5.15)$$

where T_* is the given temperature.

The direct problem of thermo-convective flow can be formulated as follows: find the velocity $\mathbf{u} = \mathbf{u}(t, \mathbf{x})$, the pressure $P = P(t, \mathbf{x})$, and the temperature $T = T(t, \mathbf{x})$ satisfying boundary value problem (5.10), (5.11) and (5.12) and initial-boundary value problem (5.13), (5.14) and (5.15). The inverse problem can be formulated in this case as follows: find the velocity, pressure, and temperature satisfying boundary value problem (5.10), (5.11) and (5.12) and the final-boundary value problem which includes Eqs. (5.13) and (5.15) and the final condition:

$$T(\vartheta, \mathbf{x}) = T_\vartheta(\mathbf{x}), \quad \mathbf{x} \in \Omega, \qquad (5.16)$$

where T_ϑ is the temperature at time $t = \vartheta$.

To solve the inverse problem by the QRV method, Ismail-Zadeh et al. (2007) considered the following regularized backward heat problem to define temperature in the past from the known temperature $T_\vartheta(\mathbf{x})$ at present time $t = \vartheta$:

$$\partial T_\beta / \partial t - \langle \mathbf{u}_\beta, \nabla T_\beta \rangle = \nabla^2 T_\beta - \beta \Lambda \left(\partial T_\beta / \partial t \right), \quad t \in [0, \vartheta], \, \mathbf{x} \in \Omega, \qquad (5.17)$$

$$\sigma_1 T_\beta + \sigma_2 \partial T_\beta / \partial \mathbf{n} = T_*, \qquad t \in (0, \vartheta), \, \mathbf{x} \in \partial\Omega, \qquad (5.18)$$

$$\sigma_1 \partial^2 T_\beta / \partial \mathbf{n}^2 + \sigma_2 \partial^3 T_\beta / \partial \mathbf{n}^3 = 0, \qquad t \in (0, \vartheta), \, \mathbf{x} \in \partial\Omega, \qquad (5.19)$$

$$T_\beta(\vartheta, \mathbf{x}) = T_\vartheta(\mathbf{x}), \qquad \mathbf{x} \in \Omega, \qquad (5.20)$$

where $\Lambda(T) = \partial^4 T / \partial x_1^4 + \partial^4 T / \partial x_2^4 + \partial^4 T / \partial x_3^4$, and the boundary value problem to determine the fluid flow:

$$\nabla P_\beta = -\nabla \cdot \left[\eta \left(T_\beta \right) \left(\nabla \mathbf{u}_\beta + \nabla \mathbf{u}_\beta^T \right) \right] + RaT_\beta \mathbf{e}, \qquad \mathbf{x} \in \Omega, \qquad (5.21)$$

$$\nabla \cdot \mathbf{u}_\beta = 0, \qquad \mathbf{x} \in \Omega, \qquad (5.22)$$

$$\langle \mathbf{u}_\beta, \mathbf{n} \rangle = 0, \quad \sigma_\beta \mathbf{n} + \langle \sigma_\beta \, \mathbf{n}, \mathbf{n} \rangle \mathbf{n} = 0, \qquad \mathbf{x} \in \partial\Omega, \qquad (5.23)$$

where $\sigma_\beta = \eta \left(\nabla \mathbf{u}_\beta + \nabla \mathbf{u}_\beta{}^T \right)$, the sign of the velocity field is changed ($\mathbf{u}_\beta$ by $-\mathbf{u}_\beta$) in Eqs. (5.17) and (5.21) to simplify the application of the total variation diminishing (TVD) method (e.g., Ismail-Zadeh and Tackley 2010; chapter 7.9) for solving (5.17), (5.18), (5.19) and (5.20). Hereinafter temperature T_ϑ is referred to as the input temperature for the problem (5.17), (5.18), (5.19), (5.20), (5.21), (5.22) and (5.23). The core of the transformation of the heat equation is the addition of a high order differential expression $\Lambda \left(\partial T_\beta / \partial t \right)$ multiplied by a small parameter $\beta > 0$. Note that Eq. (5.19) is added to the boundary conditions to properly define the regularized backward heat problem. The solution to the regularized backward heat problem is stable for $\beta > 0$, and the approximate solution to (5.17, (5.18), (5.19), (5.20), (5.21), (5.22) and (5.23) converges to the solution of (5.10), (5.11), (5.12), (5.13) and (5.14), and (5.16) in some spaces, where the conditions of well-posedness are met (Samarskii and Vabishchevich 2007). Thus, the inverse problem of thermo-convective mantle flow is reduced to determination of the velocity $\mathbf{u}_\beta = \mathbf{u}_\beta (t, \mathbf{x})$, the pressure $P_\beta = P_\beta (t, \mathbf{x})$, and the temperature $T_\beta = T_\beta (t, \mathbf{x})$ satisfying (5.17, (5.18), (5.19), (5.20), (5.21), (5.22) and (5.23).

5.3   Optimisation Problem and Numerical Approach

A maximum of the following functional is sought with respect to the regularization parameter β:

$$\delta - \left\| T \left(t = \vartheta, \cdot; T_{\beta_k} (t = 0, \cdot) \right) - \varphi \left(\cdot \right) \right\| \to \max_k, \qquad (5.24)$$

$$\beta_k = \beta_0 q^{k-1}, \ k = 1, 2, \ldots, \mathfrak{N}, \qquad (5.25)$$

where sign $\|\cdot\|$ denotes the norm in the space $L_2(\Omega)$. Here $T_k = T_{\beta_k} (t = 0, \cdot)$ is the solution to the regularized backward heat problem (5.17), (5.18) and (5.19) at $t = 0$; $T (t = \vartheta, \cdot; T_k)$ is the solution to the heat problem (5.13), (5.14) and (5.15) at the initial condition $T (t = 0, \cdot) = T_k$ at time $t = \vartheta$; φ is the known temperature at $t = \vartheta$ (the input data on the present temperature); small parameters $\beta_0 > 0$ and $0 < q < 1$ are defined below; and $\delta > 0$ is a given accuracy. When q tends to unity, the computational cost becomes large; and when q tends to zero, the optimal solution can be missed.

The prescribed accuracy δ is composed from the accuracy of the initial data and the accuracy of computations. When the input noise decreases and the accuracy of computations increases, the regularization parameter is expected to decrease. However, estimates of the initial data errors are usually inaccurate. Estimates of the computation accuracy are not always known, and when they are available, the estimates are coarse. In practical computations, it is more convenient to minimize the following functional with respect to (5.25)

$$\left\| T_{\beta_{k+1}} (t = 0, \cdot) - T_{\beta_k} (t = 0, \cdot) \right\| \to \min_k, \tag{5.26}$$

where misfit between temperatures obtained at two adjacent iterations must be compared. To implement the minimization of temperature residual (5.24), the inverse problem (5.17), (5.18), (5.19), (5.20), (5.21), (5.22) and (5.23) must be solved on the entire time interval as well as the direct problem (5.10), (5.11), (5.12), (5.13), (5.14) and (5.15) on the same time interval. This at least doubles the amount of computations. The minimization of functional (5.26) has a lower computational cost, but it does not rely on a priori information.

The numerical algorithm for solving the inverse problem of thermo-convective mantle flow using the QRV method can be described as follows. Consider a uniform temporal partition $t_n = \vartheta - \delta t \, n$ (as defined in Sect. 3.5) and prescribe some values to parameters β_0, q, and $\Re$ (e.g. $\beta_0 = 10^{-3}$, $q = 0.1$, and $\Re = 10$). According to (5.25) a sequence of the values of the regularization parameter $\{\beta_k\}$ is defined. For each value $\beta = \beta_k$ model temperature and velocity are determined in the following way.

Step 1. Given the temperature $T_\beta = T_\beta (t, \cdot)$ at $t = t_n$, the velocity $\mathbf{u}_\beta = \mathbf{u}_\beta (t_n, \cdot)$ is found by solving problem (5.21), (5.22) and (5.23). This velocity is assumed to be constant on the time interval $[t_{n+1}, t_n]$.

Step 2. Given the velocity $\mathbf{u}_\beta = \mathbf{u}_\beta (t_n, \cdot)$, the new temperature $T_\beta = T_\beta (t, \cdot)$ at $t = t_{n+1}$ is found on the time interval $[t_{n+1}, t_n]$ subject to the final condition $T_\beta = T_\beta (t_n, \cdot)$ by solving the regularized problem (5.17), (5.18), (5.19) and (5.20) backward in time.

Step 3. Upon the completion of steps 1 and 2 for all $n = 0, 1, \ldots, m$, the temperature $T_\beta = T_\beta (t_n, \cdot)$ and the velocity $\mathbf{u}_\beta = \mathbf{u}_\beta (t_n, \cdot)$ are obtained at each $t = t_n$. Based on the computed solution, find the temperature and flow velocity at each point of time interval $[0, \vartheta]$ using interpolation.

Step 4a. The direct problem (5.13), (5.14) and (5.15) is solved assuming that the initial temperature is given as $T_\beta = T_\beta (t = 0, \cdot)$, and the temperature residual (5.24) is found. If the residual does not exceed the predefined accuracy, the calculations are terminated, and the results obtained at step 3 are considered as the final ones. Otherwise, parameters β_0, q, and $\Re$ entering equation (5.25) are modified, and the calculations are continued from step 1 for new set $\{\beta_k\}$.

Step 4b. The functional (5.26) is calculated. If the residual between the solutions obtained for two adjacent regularization parameters satisfies a predefined criterion (the criterion should be defined by a user, because no a priori data are used at this step), the calculation is terminated, and the results obtained at step 3 are considered as the final ones. Otherwise, parameters β_0, q, and $\Re$ entering equation (5.25) are modified, and the calculations are continued from step 1 for new set $\{\beta_k\}$.

In a particular implementation, either step 4a or step 4b is used to terminate the computation. This algorithm allows (i) organizing a certain number of independent computational modules for various values of the regularized parameter β_k that find the solution to the regularized problem using steps 1–3 and (ii) determining *a posteriori* an acceptable result according to step 4a or step 4b.

5.4 Restoration of Mantle Plumes

To demonstrate the applicability of the QRV data assimilation method and to compare the results with those obtained by the VAR and BAD methods, Ismail-Zadeh et al. (2007) used the same forward model for mantle plume evolution as presented in Sect. 3.6.1. Figure 5.2 (panels a–d) illustrates the evolution of mantle plumes in the forward model. The state of the plumes at the "present" time (Fig. 5.2d) obtained by solving the direct problem was used as the input temperature for the inverse problem (an assimilation of the "present" temperature to the past). Note that this initial state (input temperature) is given with an error introduced by the numerical algorithm used to solve the direct problem. Figure 5.2 illustrates the states of the plumes restored by the QRV method (panels e–g) and the residual δT (see Eq. (3.13) and panel h) between the initial temperature for the forward model (Fig. 5.2a) and the temperature $\tilde{T}(\mathbf{x})$ assimilated to the same age (Fig. 5.2g). To check the stability of the algorithm, a forward model of the restored plumes is computed using the solution to the inverse problem at the time of 265 Myr ago (Fig. 5.2g) as the initial state for the forward model. The result of this run is shown in Fig. 5.2i.

To compare the accuracy of the data assimilation methods, a restoration model from the "present" time (Fig. 5.2d) to the time of 265 Myr ago was developed using the BAD method. Figure 5.2 shows the BAD model results (panels e1–g1) together with the temperature residual (panel h1) between the initial temperature (panel a) and the temperature assimilated to the same age (panel g1). The VAR method was not used to assimilate data within the time interval of more than 100 Myr (for $Ra \approx 10^6$), because proper filtering of the increasing noise is required to smooth the input data and solution (Sect. 3.7).

Figure 5.3a presents the residual $J_1(\beta) = \left\| T_0(\cdot) - T_\beta(t = t_0, \cdot; T_\vartheta) \right\|$ between the initial temperature T_0 at $t_0 = 265\,\mathrm{Myr}$ ago and the restored temperature (to the same time) obtained by solving the inverse problem with the input temperature T_ϑ. The optimal accuracy is attained at $\beta^* = \arg\min \{J_1(\beta) : \beta = \beta_k, k = 1, 2, \ldots, 10\} \approx 10^{-7}$ in the case of $r = 20$, and at $\beta^* \approx 10^{-6}$ and $\beta^* \approx 10^{-5.5}$ in the cases of the viscosity ratio $r = 200$ and $r = 1000$, respectively. Figure 5.3b illustrates the residual $J_2(\beta) = \left\| T_\beta(t_0, \cdot; T_\vartheta) - T_{\widehat{\beta}}(t_0, \cdot; T_\vartheta) \right\|$ between the reconstructed temperature at $t_0 = 265\,\mathrm{Myr}$ ago obtained for various values of β in the range $10^{-9} \leq \beta \leq 10^{-3}$ and $\widehat{\beta} = \beta/2$. These results show the choice of the optimal value of the regularization parameter using step 4b of the numerical algorithm for the QRV data assimilation. In the case of $r = 20$ the parameter $\beta^* =$

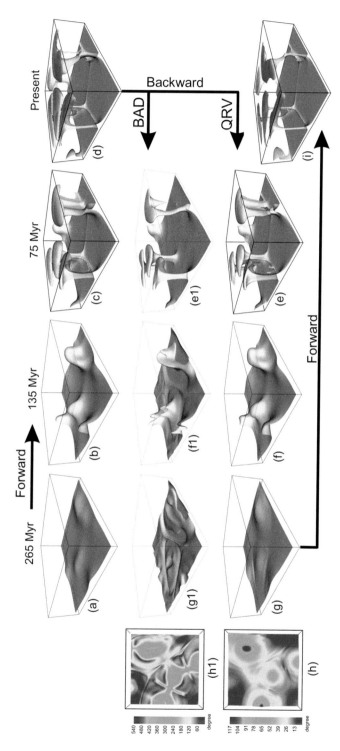

Fig. 5.2 Model of mantle plume evolution forward in time at successive times: (**a**–**d**) from 265 Myr ago to the present state of the plumes ($r = 20$). Assimilation of the mantle temperature and flow from the present state back to the geological past using the QRV (**d**–**g**; $\beta = 10^{-7}$) and BAD (**d**, **e1**–**g1**) methods. Verification of the QRV assimilation accuracy: Forward model of the plume evolution starting from the initial (restored) state of the plumes (**g**) to their present state (**i**). Temperature residuals between the initial temperature for the forward model and the temperature assimilated to the same age using the QRV and BAD methods are presented in panels (**h**) and (**h1**), respectively (After Ismail-Zadeh et al. 2007)

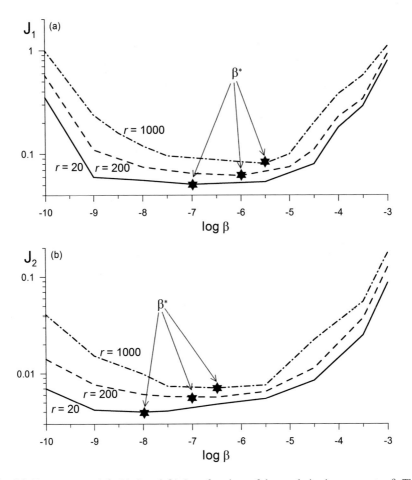

Fig. 5.3 Temperature misfit (**a**) J_1 and (**b**) J_2 as functions of the regularization parameter β. The minimum of the temperature misfit is achieved at β^*, an optimal regularization parameter. *Solid curves*: $r = 20$; *dashed curves*: $r = 200$; and *dash-dotted curves*: $r = 1000$ (After Ismail-Zadeh et al. 2007)

$\arg\min\{J_2(\beta): \beta = \beta_k,\ k = 1, 2, \ldots, 12\} \approx 10^{-8}$ provides the optimal accuracy for the solution; in the cases of $r = 200$ and $r = 1000$ the optimal accuracy is achieved at $\beta^* \approx 10^{-7}$ and $\beta^* \approx 10^{-6.5}$, respectively. Comparison of the temperature residuals for three values of the viscosity ratio r indicates that the residuals become larger as the viscosity ratio increases. The numerical experiments show that the algorithm for solving the inverse problem performs well when the regularization parameter is in the range $10^{-8} \leq \beta \leq 10^{-6}$. For greater values, the solution of the inverse problem retains the stability but is less accurate. The numerical procedure becomes unstable at $\beta < 10^{-9}$, and the computations must be stopped.

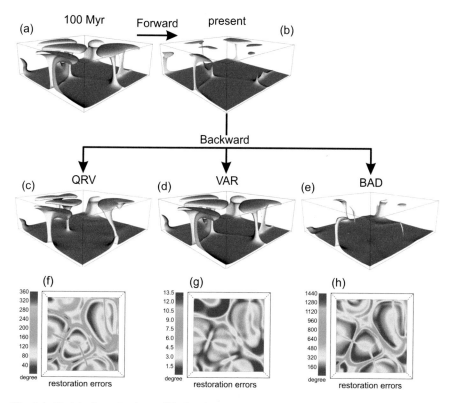

Fig. 5.4 Model of mantle plume diffusion forward in time (**a** and **b**; $r = 20$). Assimilation of the mantle temperature and flow to the time of 100 Myr ago and temperature residuals between the present temperature model (**b**) and the temperature assimilated to the same age, using the QRV (**c** and **f**; $\beta = 10^{-7}$), VAR (**d** and **g**), and BAD (**e** and **h**) methods, respectively (After Ismail-Zadeh et al. 2007)

To compare how the techniques for data assimilation can restore the prominent state of the thermal plumes in the past from their 'present' weak state, a forward model was initially developed from the prominent state of the plumes (Fig. 5.4a) to their diffusive state in 100 Myr (Fig. 5.4b) using $50 \times 50 \times 50$ finite rectangular elements to approximate the vector velocity potential and a finite difference grid $148 \times 148 \times 148$ for approximation of temperature, velocity, and viscosity. All other parameters of the model are the same.

The VAR method (Fig. 5.4d, g) provides the best performance for the diffused plume restoration. The BAD method (Fig. 5.4e, h) cannot restore the diffused parts of the plumes, because temperature is only advected backward in time. The QRV method (Fig. 5.4c, f) restores the diffused thermal plumes, meanwhile the restoration results are not so perfect as in the case of VAR method (compare temperature residuals in Fig. 5.4, panels f and g). Although the accuracy of the QRV data assimilation is lower compared to the VAR data assimilation, the QRV method does not require any additional smoothing of the input data and filtering of temperature noise as the VAR method does.

5.5   Restoration of Descending Lithosphere Evolution

5.5.1   The Vrancea Seismicity and the Relic Descending Slab

Repeated large intermediate-depth earthquakes in the south-eastern (SE) Carpathians (the Vrancea region) cause destruction in Bucharest, the capital city of Romania, and shake central and eastern European cities several hundred kilometres away from the hypocentres of the events. We refer readers to a comprehensive review by Ismail-Zadeh et al. (2012) on geology, tectonics, geodynamics, geodesy, seismic and geoelectric studies in the region as well as on the modelling efforts in developing regional density/gravity and thermal structure, stress and strain, lithospheric deformation and earthquake simulation, seismic hazard and earthquake forecasting in Vrancea. Here we briefly describe the region, intermediate depth seismicity beneath Vrancea, and its association with a high velocity body in the uppermost mantle revealed by seismic tomography studies.

The earthquake-prone Vrancea region (Fig. 5.5) is bounded to the north and northeast by the Eastern European platform (EEP), to the east by the Scythian platform (SCP), to the south-east by the North Dobrogea orogen (DOB), to the south and south-west by the Moesian platform (MOP), and to the north-west by the Transylvanian basin (TRB). The epicentres of the sub-crustal earthquakes in the Vrancea region are concentrated within a very small seismogenic volume about $70 \times 30 \, \mathrm{km^2}$ in planform and between depths of about 70 and 180 km. Below this depth the seismicity ends abruptly: one seismic event at 220 km depth is an exception (Oncescu and Bonjer 1997).

The 1940 $M_W = 7.7$ earthquake gave rise to the development of a number of geodynamic models for this region. McKenzie (1972) suggested that this seismicity is associated with a relic slab sinking in the mantle and now overlain by continental crust. The 1977 large earthquake and later the 1986 and 1990 earthquakes again raised questions about the nature of the earthquakes. A seismic gap at depths of 40–70 km beneath Vrancea led to the assumption that the lithospheric slab had already detached from the continental crust (Fuchs et al. 1979). Oncescu (1984) proposed that the intermediate-depth events are generated in a zone that separates the sinking slab from the neighbouring immobile part of the lithosphere rather than in the sinking slab itself. Linzer (1996) explained the nearly vertical position of the Vrancea slab as the final rollback stage of a small fragment of oceanic lithosphere. Various types of slab detachment or delamination (e.g. Wortel and Spakman 2000; Sperner et al. 2001) have been proposed to explain the present-day seismic images of the descending slab. Cloetingh et al. (2004) argued in favour of the complex configuration of the underthrusted lithosphere and its thermo-mechanical age as primary factors in the behaviour of the descending slab after continental collision.

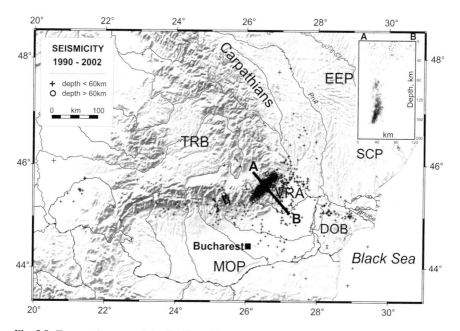

Fig. 5.5 Topography map of the SE-Carpathians and epicentres of Vrancea earthquakes (magnitude $\geq$ 3). *Upper right panel* presents hypocentres of the same earthquakes projected onto the NW-SE vertical plane AB. *DOB* Dobrogea orogen; *EEP* Eastern European platform; *MOP* Moesian platform; *SCP* Scythian platform; *TRB* Transylvanian basin; and *VRA* Vrancea (After Ismail-Zadeh et al. 2008)

The origin of the descending lithosphere in the region, i.e. whether the Vrancea slab is oceanic or continental, is still under debate.

The Neogene to Late Miocene evolution of the Carpathian region is mainly driven by the north-eastward, later eastward and south-eastward roll-back or slab retreat (Royden 1988; Sperner et al. 2001) into a Carpathians embayment, consisting of the last remnants of an oceanic or thinned continental domain attached to the European continent (e.g. Balla 1987). When the mechanically strong East-European and Scythian platforms started to enter the subduction zone, the buoyancy forces of the thick continental crust exceeded the slab pull forces and convergence stopped after only a short period of continental thrusting (e.g., Tarapoanca et al. 2004). Continental convergence in the SE-Carpathians ceased about 11 Ma (e.g., Csontos et al. 1992), and after that the lithospheric slab descended beneath the Vrancea region due to gravity. The hydrostatic buoyancy forces promote the sinking of the slab, but viscous and frictional forces resist the descent. The combination of these forces produces shear stresses at intermediate depths that are high enough to cause earthquakes (Ismail-Zadeh et al. 2000, 2005a, 2010).

Here we present a quantitative model of the thermal evolution of the descending slab in the SE-Carpathians suggested by Ismail-Zadeh et al. (2008). The model is based on assimilation of present crust/mantle temperature and flow in the geological

past using the QRV method. Mantle thermal structures are restored and analysed in
the context of modern regional geodynamics.

5.5.2 Input Data: Seismic Temperature Model

Temperature is a key physical parameter controlling the density and rheology
of the Earth's material and hence crustal and mantle dynamics. Besides direct
measurements of temperature in boreholes in the shallow portion of the crust, there
are no direct measurements of deep crustal and mantle temperatures, and therefore
the temperatures must be estimated indirectly from seismic wave anomalies,
geochemical data, and surface heat flow observations.

Ismail-Zadeh et al. (2005a, 2008) developed a model of the present crustal
and mantle temperature beneath the SE-Carpathians using the most recent high-
resolution seismic tomography image (map of the anomalies of P-wave velocities)
of the lithosphere and asthenosphere in the region (Martin et al. 2005, 2006). The
tomography image shows a high velocity body beneath the Vrancea region and
the Moesian platform interpreted as the subducted lithospheric slab (Martin et al.
2006). The seismic tomographic model of the region consists of eight horizontal
layers of different thickness (15 km up to 70 km) starting from the depth of 35 km
and extending down to a depth of 440 km. Each layer of about 1000×1000 km^2
is subdivided horizontally into 16×16-km^2 blocks. To restrict numerical errors in
our data assimilation, the velocity anomaly data are smoothed between the blocks
and the layers using a spline interpolation. Ismail-Zadeh et al. (2005a) converted
seismic wave velocity anomalies into temperature considering the effects of mantle
composition, anelasticity, and partial melting on seismic velocities. The temperature
in the crust is constrained by measurements of surface heat flux corrected for
paleoclimate changes and for the effects of sedimentation.

Depth slices of the present temperature model are illustrated in Fig. 5.6. The
pattern of resulting mantle temperature anomalies (predicted temperature minus
background temperature) is similar to the pattern of observed P-wave velocity
anomalies (Martin et al. 2006), but not an exact copy because of the nonlinear
inversion of the seismic anomalies to temperature. The low temperatures are
associated with the high-velocity body beneath the Vrancea region (VRA) and the
East European platform (EEP) and are already visible at depths of 50 km. The slab
image becomes clear at 70–110 km depth as a NE-SW oriented cold anomaly. With
increasing depth (110–200 km depth) the thermal image of the slab broadens in
NW–SE direction. The orientation of the cold body changes from NE–SW to N–S
below the depth of 200 km. The slab extends down to 280–320 km depth beneath the
Vrancea region itself. A cold anomaly beneath the Transylvanian Basin is estimated
at depths of 370–440 km. According to Wortel and Spakman (2000) and Martin et
al. (2006) this cold material can be interpreted as a remnant of subducted lithosphere
detached during the Miocene along the Carpathian Arc and residing within the upper
mantle transition zone. High temperatures are predicted beneath the Transylvanian

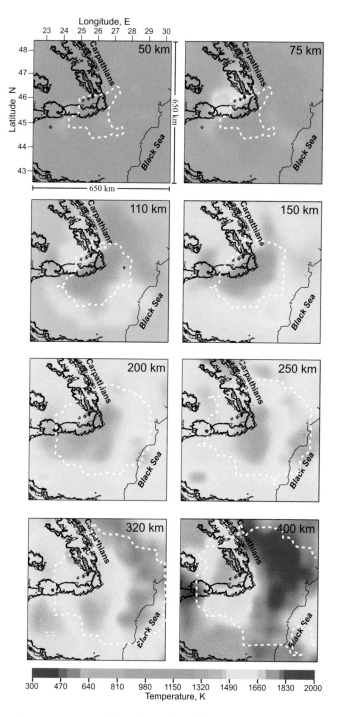

Fig. 5.6 Present temperature model as the result of the inversion of the P-wave velocity model. Theoretically well-resolved regions are bounded by dashed line (Martin et al. 2006). Each slice presents a part of the horizontal section of the model domain Ω corresponding to $[x_1 = 177.5 \text{ km}, x_1 = 825.5 \text{ km}] \times [x_2 = 177.5 \text{ km}, x_2 = 825.5 \text{ km}]$, and the isolines present the surface topography (also in Fig. 5.7) (After Ismail-Zadeh et al. 2008)

Basin (TRB) at about 70–110 km depth. Two other high temperature regions are found at 110–150 km depth below the Moesian platform (MOP) and deeper than 200 km under the EEP and the Dobrogea orogen (DOB), which might be correlated with the regional lithosphere/asthenosphere boundary.

5.5.3   Data Assimilation

To minimize boundary effects, the studied region (650×650 km^2 and 440 km deep, see Fig. 5.6) has been bordered horizontally by 200 km area and extended vertically to the depth of 670 km. Therefore, a rectangular domain $\Omega = [0, l_1 = 1050$ km$] \times [0, l_2 = 1050$ km$] \times [0, h = 670$ km$]$ is considered for assimilation of present temperature and mantle flow beneath the SE-Carpathians.

Our ability to reverse mantle flow is limited by our knowledge of past movements in the region, which are well constrained only in some cases. In reality, the Earth's crust and lithospheric mantle are driven by mantle convection and the gravitational pull of dense descending slabs. However, when a numerical model is constructed for a particular region, external lateral forces can influence the regional crustal and uppermost mantle movements. Yet in order to make useful predictions that can be tested geologically, a time-dependent numerical model should include the history of surface motions. Since this is not currently achievable in a dynamical way, it is necessary to prescribe surface motions using velocity boundary conditions.

The simulations are performed backward in time for a period of 22 Myr. Perfect slip conditions are assumed at the vertical and lower boundaries of the model domain. For the first 11 Myr (starting from the present time), when the rates of continental convergence were insignificant (e.g., Csontos et al. 1992), no velocity is imposed at the surface, and the conditions at the upper boundary are free slip. The north-westward velocity is imposed in the portion of the upper model boundary (Fig. 5.7a) for the time interval from 11 Myr to 16 Myr and the westward velocity in the same portion of the boundary (Fig. 5.7b) for the interval from 16 Myr to 22 Myr. The velocities are consistent with the direction and rates of the regional convergence in the Early and Middle Miocene (e.g., Fügenschuh and Schmid 2005). The effect of the surface loading due to the Carpathian Mountains is not considered, because this loading would have not major influence on the dynamics of the region (as was shown in two-dimensional models of the Vrancea slab evolution; Ismail-Zadeh et al. 2005b).

The heat flux through the vertical boundaries of the model domain Ω is set to zero. The upper and lower boundaries are assumed to be isothermal surfaces. The present temperature above 440 km depth is derived from the seismic velocity anomalies and heat flow data. The adiabatic geotherm for potential temperature 1750 K (Katsura et al. 2004) was used to define the present temperature below 440 km (where seismic tomography data are not available). Relevant equations (5.10), (5.11), (5.12), (5.13), (5.14), (5.15), (5.16), (5.17), (5.18), (5.19), (5.20), (5.21), (5.22) and (5.23) are solved numerically.

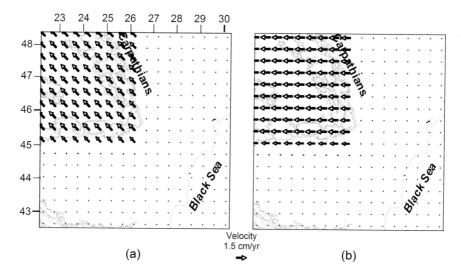

Fig. 5.7 Surface velocity imposed on the part of the upper boundary of the model domain in data assimilation modelling for the time interval from 11 Myr to 16 Myr ago (**a**) and for that from 16 Myr to 22 Myr ago (**b**) (After Ismail-Zadeh et al. 2008)

To estimate the accuracy of the results of data assimilation, the temperature and mantle flow restored to the time of 22 Myr ago were employed as the initial condition for a model of the slab evolution forward in time; the model was run to the present; and the temperature residual (the difference between the present temperature and that predicted by the forward model with the restored temperature as an initial temperature distribution) was analysed subsequently. The maximum temperature residual does not exceed 50 degrees.

A sensitivity analysis was performed to understand how stable is the numerical solution to small perturbations of input (present) temperatures. The present temperature model was perturbed randomly by 0.5–2 % and then assimilated to the past to find the initial temperature. A misfit between the initial temperatures related to the perturbed and unperturbed present temperature is rather small (2–4 %) which proves that the solution is stable. The numerical models, with a spatial resolution of 7 km × 7 km × 5 km, were run on parallel computers. The accuracy of the numerical solutions has been verified by several tests, including grid and total mass changes (Ismail-Zadeh et al. 2001).

5.5.4 Model Results

The results related to assimilation of the present temperature model beneath the SE-Carpathians into Miocene times are discussed here. Although there is some evidence that the lithospheric slab was already partly subducted some 75 Myr ago (e.g. Sandulescu 1988), the assimilation interval was restricted to the Miocene, because the pre-Miocene evolution of the descending slab, as well as the regional horizontal movements, are poorly known. Incorporation of insufficiently accurate data into the assimilation model could result in incorrect scenarios of mantle and lithosphere dynamics in the region. Moreover, to restore the history of pre-Miocene slab subduction, a high resolution seismic tomography image of the deeper mantle is required (the present image is restricted to the depth of 440 km).

Early Miocene subduction beneath the Carpathian arc and the subsequent gentle continental collision transported cold and dense lithospheric material into the hotter mantle. Figure 5.8 presents the 3-D thermal image of the slab and pattern of contemporary flow induced by the descending slab. Note that the direction of the flow is reversed, because the problem is solved backward in time: cold slab move upward during numerical backward modelling. The 3-D flow is rather complicated: toroidal (in horizontal planes) flow at depths between about 100 to 200 km coexists with poloidal (in vertical planes) flow.

The relatively cold region seen at depths of 40–230 km can be interpreted as the earlier evolutionary stages of the lithospheric slab. Since active subduction of the lithospheric slab in the region ended in Late Miocene time and earlier rates of convergence were low before it, Ismail-Zadeh et al. (2008) argue that the cold slab, descending slowly at these depths, has been warmed up, and its thermal shape has faded due to heat diffusion. Thermal conduction in the shallow Earth (where viscosity is high) plays a significant part in heat transfer compared to thermal convection. The deeper we look in the region, the larger are the effects of thermal advection compared to conduction: the lithosphere has moved upwards to the place where it had been in Miocene times (Fig. 5.9). Below 230 km depth the thermal roots of the cold slab are clearly visible in the present temperature model, but they are less prominent in the assimilated temperature images, because the slab did not reach these depths in Miocene times.

The geometry of the restored slab clearly shows two parts of the sinking body (Fig. 5.9). The NW-SE oriented part of the body is located in the vicinity of the boundary between the EEP and Scythian platform (SCP) and may be a relic of cold lithosphere that has travelled eastward. Another part has a NE-SW orientation and is associated with the present descending slab. An interesting geometrical feature of the restored slab is its curvature beneath the SE-Carpathians. In Miocene times the slab had a concave surface confirming the curvature of the Carpathian arc down to depths of about 60 km. At greater depths the slab changed its shape to that of a convex surface and split into two parts at a depth of about 200 km. Although such a change in slab curvature is visible neither in the model of the present temperature nor in the seismic tomography image most likely because of slab warming and heat diffusion, the convex shape of the slab is likely to be preserved at the present time. Ismail-Zadeh et al. (2008) proposed that this change in the geometry of the descending slab can cause stress localization due to slab bending and subsequent stress release resulting in earthquakes, which occur at depths of 70–180 km in the region.

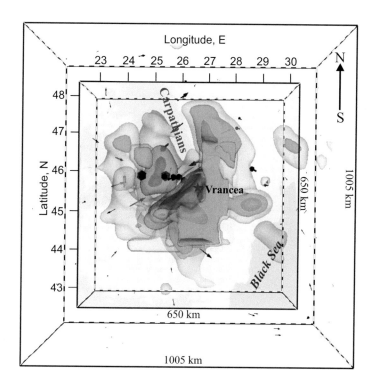

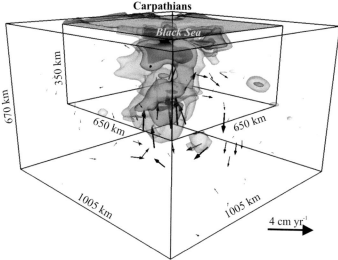

Fig. 5.8 3-D thermal shape of the Vrancea slab and contemporary flow induced by the descending slab beneath the SE-Carpathians. *Upper panel*: top view. *Lower panel*: side view from the SE toward NW. *Arrows* illustrate the direction and magnitude of the flow. The marked sub-domain of the model domain presents the region around the Vrancea shown in Fig. 5.9. The surfaces marked by *blue, dark cyan,* and *light cyan* illustrate the surfaces of 0.07, 0.14, and 0.21 temperature anomaly δT, respectively, where $\delta T = (T_{hav} - T)/T_{hav}$, and T_{hav} is the horizontally averaged temperature. The top surface presents the topography, and the *red star* marks the location of the intermediate-depth earthquakes (After Ismail-Zadeh et al. 2008)

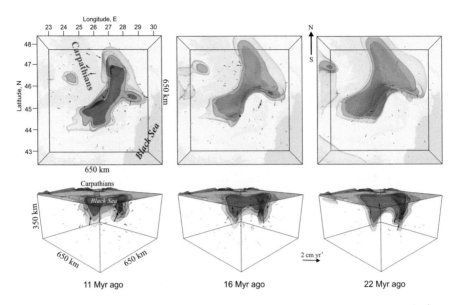

Fig. 5.9 Snapshots of the 3-D thermal shape of the Vrancea slab and pattern of mantle flow beneath the SE-Carpathians in the Miocene times (See Fig. 5.8 for other notations. After Ismail-Zadeh et al. 2008)

Moreover, the north-south (NS)-oriented cold material visible at the depths of 230–320 km (Fig. 5.6) does not appear as a separate (from the NE-SW-oriented slab) body in the models of Miocene time. Instead, it looks more like two differently oriented branches of the SW-end of the slab at 60–130 km depth (visible in Fig. 5.9). Therefore, the results of the assimilation of the present temperature model to Miocene time provide a plausible explanation for the change in the spatial orientation of the slab from NE-SW to NS beneath 200 km observed in the seismic tomography image (Martin et al. 2006). The slab bending might be related to a complex interaction between two parts of the sinking body and the surrounding mantle. The sinking body displaces the mantle, which, in its turn, forces the slab to deform due to corner (toroidal) flows different within each of two sub-regions (to NW and to SE from the present descending slab). Also, the curvature of the descending slab can be influenced by slab heterogeneities due to variations in its thickness and viscosity (Cloetingh et al. 2004; Morra et al. 2006).

Martin et al. (2006) interpret the negative velocity anomalies NW of the present slab at depths between 70 and 110 km (see the relevant temperature slices in Fig. 5.6) as a shallow asthenospheric upwelling associated with possible slab rollback. Also, they mention partial melting as an additional contribution to the reduction of seismic velocities at these depths. The results of our assimilation show that the descending slab is surrounded by a border of hotter rocks at depths down to about 250 km. The rocks could be heated due to partial melting as a result of slab dehydration. Although the effects of slab dehydration or partial melting were not considered in the modelling, the numerical results support the hypothesis of dehydration of the descending lithosphere and its partial melting as the source of reduction of seismic velocities at these depths and probably deeper (see temperature slices at the depths of 130–220 km). Alternatively, the hot anomalies beneath the Transylvanian basin and partly beneath the Moesian platform could be dragged down by the descending slab since the Miocene times, and therefore, the slab was surrounded by the hotter rocks. Using numerical experiments Honda et al. (2007) showed recently how the lithospheric plate subducting beneath the Honshu island in Japan dragged down a hot anomaly adjacent to the plate. Some areas of high temperature at depths below 280 km can be associated with mantle upwelling in the region. High-temperature anomalies are not clearly visible in the restored temperatures at these depths, because the upwelling was likely not active in Miocene times.

5.5.5 Model Limitations and Uncertainties

There is a major physical limitation of the restoration of mantle structures. If a thermal feature created, let us say, hundreds million years ago has completely diffused away by the present, it is impossible to restore the feature, which was more prominent in the past. The time to which a present thermal structure in the upper mantle can be restored should be restricted by the characteristic thermal diffusion time, the time when the temperatures of the evolved structure and the ambient mantle are nearly indistinguishable (Ismail-Zadeh et al. 2004). The time (t) for restoration of seismic thermal structures depends on depth (d) of seismic tomography images and can be roughly estimated as $t = d/v$, where v is the average vertical velocity of mantle flow. For example, the time for restoration of the Vrancea slab evolution in the studied models should be less than about 80 Myr, considering $d = 400$ km and $v \approx 0.5$ cm yr^{-1}.

Other sources of uncertainty in the modelling of mantle temperature in the SE-Carpathians come from the choice of mantle composition (e.g., Szabó et al. 2004), the seismic attenuation model (Popa et al. 2005; Weidle et al. 2007), and poor knowledge of the presence of water at mantle depths. The drop of electrical resistivity below 1 Ω m (Stanica and Stanica 1993) can be an indicator of the presence of fluids (due to dehydration of mantle rocks) below the SE-Carpathians; however, the information is very limited and cannot be used in quantitative modelling. Viscosity is an important physical parameter in numerical modelling of mantle dynamics, because it influences the stress state and results in strengthening or weakening of Earth's material. Though it is the least-known physical parameter of the model, the viscosity of the Vrancea slab was constrained by observations of the regional strain rates (Ismail-Zadeh et al. 2005a).

The geometry of the mantle structures changes with time, diminishing the degree of surface curvature of the structures. Heat conduction smoothes the complex thermal surfaces of mantle bodies with time. Present seismic tomography images of mantle structures do not allow definition of the sharp shapes of these structures. Assimilation of mantle temperature and flow to the geological past instead provides a quantitative tool to restore thermal shapes of prominent structures in the past from their diffusive shapes at present. High-resolution experiments on seismic wave attenuation, improved knowledge of crustal and mantle mineral composition, accurate GPS measurements of regional movements, and precise geological paleo-reconstructions of crustal movements will assist to refine the present models and our knowledge of the regional thermal evolutions. The basic knowledge gained from the case studies is the dynamics of the Earth interior in the past, which could result in its present dynamics.

References

Balla Z (1987) Tertiary paleomagnetic data for the Carpatho-Pannonian region in the light of Miocene rotation kinematics. Tectonophysics 139:67–98

Bubnov VA (1976) Wave concepts in the theory of heat. Int J Heat Mass Transf 19:175–184

Bubnov VA (1981) Remarks on wave solutions of the nonlinear heat-conduction equation. J Eng Phys Thermophys 40(5):565–571

Cattaneo C (1958) Sur une forme de l'equation de la chaleur elinant le paradox d'une propagation instantance. Comptes Rendus 247:431–433

Cloetingh SAPL, Burov E, Matenco L, Toussaint G, Bertotti G, Andriessen PAM, Wortel MJR, Spakman W (2004) Thermo-mechanical controls on the model of continental collision in the SE Carpathians (Romania). Earth Planet Sci Lett 218:57–76

Csontos L, Nagymarosy A, Horvath F, Kovac M (1992) Tertiary evolution of the intra-Carpathian area; a model. Tectonophysics 208:221–241

Fuchs K, Bonjer K, Bock G, Cornea I, Radu C, Enescu D, Jianu D, Nourescu A, Merkler G, Moldoveanu T, Tudorache G (1979) The Romanian earthquake of March 4, 1977. II. Aftershocks and migration of seismic activity. Tectonophysics 53:225–247

Fügenschuh B, Schmid SM (2005) Age and significance of core complex formation in a very curved orogen: Evidence from fission track studies in the South Carpathians (Romania). Tectonophysics 404:33–53

Honda S, Morishige M, Orihashi Y (2007) Sinking hot anomaly trapped at the 410 km discontinuity near the Honshu subduction zone. Jpn Earth Planet Sci Lett 261:565–577

Ismail-Zadeh A, Tackley P (2010) Computational methods for geodynamics. Cambridge University Press, Cambridge

Ismail-Zadeh A, Aoudia A, Panza GF (2010) Three-dimensional numerical modeling of contemporary mantle flow and tectonic stress beneath the Central Mediterranean. Tectonophysics 482:226–236

Ismail-Zadeh AT, Panza GF, Naimark BM (2000) Stress in the descending relic slab beneath the Vrancea region, Romania. Pure Appl Geophys 157:111–130

Ismail-Zadeh AT, Korotkii AI, Naimark BM, Tsepelev IA (2001) Numerical modelling of three-dimensional viscous flow with gravitational and thermal effects. Comput Math Math Phys 41(9):1331–1345

Ismail-Zadeh A, Schubert G, Tsepelev I, Korotkii A (2004) Inverse problem of thermal convection: numerical approach and application to mantle plume restoration. Phys Earth Planet Inter 145:99–114

Ismail-Zadeh A, Mueller B, Schubert G (2005a) Three-dimensional modeling of present-day tectonic stress beneath the earthquake-prone southeastern Carpathians based on integrated analysis of seismic, heat flow, and gravity observations. Phys Earth Planet Inter 149:81–98

Ismail-Zadeh A, Mueller B, Wenzel F (2005b) Modelling of descending slab evolution beneath the SE-Carpathians: implications for seismicity. In: Wenzel F (ed) Perspectives in modern seismology, lecture notes in earth sciences, vol 105. Springer, Heidelberg, pp 205–226

Ismail-Zadeh A, Korotkii A, Schubert G, Tsepelev I (2007) Quasi-reversibility method for data assimilation in models of mantle dynamics. Geophys J Int 170:1381–1398

Ismail-Zadeh A, Schubert G, Tsepelev I, Korotkii A (2008) Thermal evolution and geometry of the descending lithosphere beneath the SE-Carpathians: an insight from the past. Earth Planet Sci Lett 273:68–79

Ismail-Zadeh A, Matenco L, Radulian M, Cloetingh S, Panza G (2012) Geodynamic and intermediate-depth seismicity in Vrancea (the south-eastern Carpathians): current state-of-the-Art. Tectonophysics 530–531:50–79

Katsura T, Yamada H, Nishikawa O, Song M, Kubo A, Shinmei T, Yokoshi S, Aizawa Y, Yoshino T, Walter MJ, Ito E, Funakoshi K (2004) Olivine-wadsleyite transition in the system (Mg, Fe)$_2$SiO$_4$. J Geophys Res 109:B02209. doi:10.1029/2003JB002438

Kirsch A (1996) An introduction to the mathematical theory of inverse problems. Springer, New York

Lattes R, Lions JL (1969) The method of quasi-reversibility: applications to partial differential equations. Elsevier, New York

Linzer H-G (1996) Kinematics of retreating subduction along the Carpathian arc, Romania. Geology 24:167–170

Martin M, Ritter JRR, The CALIXTO Working Group (2005) High-resolution teleseismic body-wave tomography beneath SE Romania – I. Implications for three-dimensional versus one-dimensional crustal correction strategies with a new crustal velocity model. Geophys J Int 162:448–460

Martin M, Wenzel F, The CALIXTO Working Group (2006) High-resolution teleseismic body wave tomography beneath SE-Romania – II. Imaging of a slab detachment scenario. Geophys J Int 164:579–595

McKenzie DP (1972) Active tectonics of the Mediterranean region. Geophys J Roy Astron Soc 30:109–185

Morra G, Regenauer-Lieb K, Giardini D (2006) Curvature of oceanic arcs. Geology 34:877–880

Morse PM, Feshbach H (1953) Methods of theoretical physics. McGraw-Hill, New York

Oncescu MC (1984) Deep structure of the Vrancea region, Romania, inferred from simultaneous inversion for hypocentres and 3-D velocity structure. Ann Geophys 2:23–28

Oncescu MC, Bonjer KP (1997) A note on the depth recurrence and strain release of large Vrancea earthquakes. Tectonophysics 272:291–302

Popa M, Radulian M, Grecu B, Popescu E, Placinta AO (2005) Attenuation in Southeastern Carpathians area: result of upper mantle inhomogeneity. Tectonophysics 410:235–249

Royden LH (1988) Late Cenozoic tectonics of the Pannonian basin system. Am Assoc Petr Geol Mem 45:27–48

Samarskii AA, Vabishchevich PN (2007) Numerical methods for solving inverse problems of mathematical physics. De Groyter, Berlin

Sandulescu M (1988) Cenozoic tectonic history of the Carpathians. Am Assoc Petr Geol Mem 45:17–25

Sperner B, Lorenz F, Bonjer K, Hettel S, Müller B, Wenzel F (2001) Slab break-off – abrupt cut or gradual detachment? New insights from the Vrancea region (SE Carpathians, Romania). Terra Nova 13:172–179

Stanica D, Stanica M (1993) An electrical resistivity lithospheric model in the Carpathian orogen from Romania. Phys Earth Planet Inter 81:99–105

Szabó C, Falus G, Zajacz Z, Kovacs I, Bali E (2004) Composition and evolution of lithosphere beneath the Carpathian-Pannonian Region: a review. Tectonophysics 393:119–137

Tarapoanca M, Carcia-Catellanos D, Bertotti G, Matenco L, Cloetingh SAPL, Dinu C (2004) Role of the 3-D distributions of load and lithospheric strength in orogenic arcs: polystage subsidence in the Carpathians foredeep. Earth Planet Sci Lett 221:163–180

Vernotte P (1958) Les paradoxes de la theorie continue de l'equation de la chaleur. Comptes
   Rendus 246:3154–3155
Weidle C, Wenzel F, Ismail-Zadeh A (2007) t* – an unsuitable parameter for anelastic attenuation
   in the Eastern Carpathians. Geophys J Int 170:1139–1150
Wortel MJR, Spakman W (2000) Subduction and slab detachment in the Mediterranean–
   Carpathian region. Science 290:1910–1917
Yu N, Imatani S, Inoue T (2004) Characteristics of temperature field due to pulsed heat input
   calculated by non-Fourier heat conduction hypothesis. JSME Int J Ser A 47(4):574–580

Chapter 6
Application of the QRV Method to Modelling of Plate Subduction

Abstract This chapter presents the application of the QRV method to dynamic restoration of the thermal state of the mantle beneath the Japanese islands and their surroundings. The geodynamic restoration for the last 40 million years is based on the assimilation of the present temperature inferred from seismic tomography, and the present plate movement derived from geodetic observations, paleogeographic and paleomagnetic plate reconstructions.

Keywords Quasi-reversibility • Subduction • Pacific plate • Lithospheric breaches • Tears • Japan Sea • Hot upwelling • Numerical modelling

6.1 Plate Subduction Beneath the Japanese Islands

An interaction of the Pacific, Okhotsk, Eurasian, and Philippine Sea lithosphere plates with the deeper mantle around the Japanese islands (Fig. 6.1) is complicated by active subduction of the plates (Fukao et al. 2001; Furumura and Kennett 2005) and back-arc spreading (Jolivet et al. 1994), which cannot be understood by the plate kinematics only. The Pacific plate subducts under the Okhotsk and the Philippine Sea plates with the relative speed of about 9 cm yr^{-1} and 5 cm yr^{-1}, respectively, whereas the Philippine Sea plate subducts under the Eurasian plate with the relative speed of about 5 cm yr^{-1} (Drewes 2009). Back arcs of these subduction zones are also known as the site of active spreading in the past and recent as inferred from both the geophysical and geological surveys (Jolivet et al. 1994). Elucidating a cause of the Japan Sea back-arc opening is one of scientific challenges in geosciences. Although its kinematic description and/or qualitative images of dynamics are fairly well understood, quantitative interpretations based on sound physical and other principles are yet missing.

P-wave seismic tomography of the mantle beneath the subducting Pacific plate near the Japanese islands revealed a low velocity region extending oceanward at depths around the 410-km seismic discontinuity, and this low velocity anomaly region was interpreted as a zone with an excess temperature of 200 K and the associated fractional melt of less than 1 % (Obayashi et al. 2006). To clarify the origin of the hot temperature anomaly beneath the Pacific plate and its implication

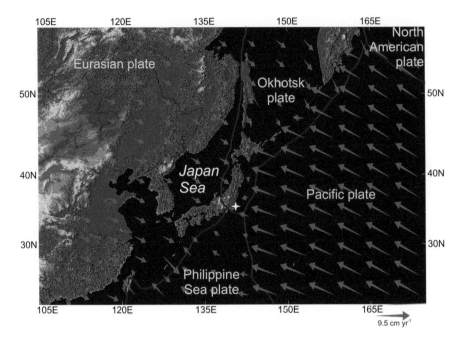

Fig. 6.1 Topography map of the Japanese Islands and surroundings. The plate motions and deformations are presented by *arrows*. The rate of the motions is determined from geodetic data and for the Philippine Sea plate from the PB2002 model (see the text). The *white star* indicates the place of sampling for geochemical analysis (Hanyu et al. 2006) (After Ismail-Zadeh et al. 2013)

for back-arc basin evolution, Ismail-Zadeh et al. (2013) studied the mantle evolution beneath the Japanese islands and their surroundings based on the assimilation of temperature inferred from seismic tomography (Fukao et al. 2001), the present movements derived from geodetic observations (Drewes 2009), and the past plate motion inferred from paleogeographic and paleomagnetic plate reconstructions (Seno and Maruyama 1984; Northrup et al. 1995; Hall 2002; Yamazaki et al. 2010). In this chapter we present and discuss the model by Ismail-Zadeh et al. (2013).

6.2   Mathematical Statement

In the three-dimensional (3-D) rectangular domain $\overline{\Omega} = [0, x_1 = l_1 = 4000 \text{ km}] \times [0, x_2 = l_2 = 4000 \text{ km}] \times [0, x_3 = h = 800 \text{ km}]$ and for time interval $t \in [0, \vartheta]$, the regularized Stokes, the incompressibility, and the backward heat balance equations are solved using the QRV method (chapter 5; Ismail-Zadeh et al. 2007) and the extended Boussinesq approximation (Christensen and Yuen 1985):

$$-\nabla P + \nabla \cdot \left(\eta \left[\nabla \mathbf{u} + (\nabla \mathbf{u})^T \right] \right) = \left(\mathbf{E} + \varsigma \nabla^2 \right)^{-1} \left[RaT - a_1 La\Phi_1 (\pi_1) - a_2 La\Phi_2 (\pi_2) \right] \mathbf{e}, \tag{6.1}$$

$$\nabla \cdot \mathbf{u} = 0, \tag{6.2}$$

$$\frac{\partial}{\partial t} \left(\mathbf{E} + \beta \nabla^2 \right)^2 T - \mathbf{u} \cdot \nabla T - A^{-1} B \, Di^* Ra \, u_3 \, T = -A^{-1} \left(-\nabla^2 T + Di^* \eta \sum_{i,j=1}^{3} \left(e_{ij} \right)^2 \right) \tag{6.3}$$

with appropriate boundary and initial conditions (see below). Here

$$A = \left[1 + \left(a_1 \frac{2}{w_1} \left(\Phi_1 - \Phi_1{}^2 \right) \overline{\gamma}_1^2 + a_2 \frac{2}{w_2} \left(\Phi_2 - \Phi_2{}^2 \right) \overline{\gamma}_2^2 \right) Di^* \, La \, T \right] > 0,$$

$$B = \left[1 + \frac{La}{Ra} \left(a_1 \frac{2}{w_1} \left(\Phi_1 - \Phi_1{}^2 \right) \overline{\gamma}_1 + a_2 \frac{2}{w_2} \left(\Phi_2 - \Phi_2{}^2 \right) \overline{\gamma}_2 \right) \right],$$

$$\Phi_i = \frac{1}{2} \left[1 + \tanh \frac{\pi_i}{w_i} \right], \quad \pi_i = z_i - x_3 - \overline{\gamma}_i (T - T_i), \quad i = 1, \, 2,$$

$\mathbf{x} = (x_1, x_2, x_3)$, $\mathbf{u} = (u_1, u_2, u_3)$, t, T, P, and η are the dimensionless Cartesian coordinates, velocity, time, temperature, pressure, and viscosity, respectively; ϑ is the present time; $e_{ij} (\mathbf{u}) = \{ \partial u_i / \partial x_j + \partial u_j / \partial x_i \}$ is the strain rate tensor; $\mathbf{e} = (0, 0, 1)$ is the unit vector; ∇ is the gradient operator; and $\mathbf{E}$ is the unit operator. With regard to the phase changes around 410 km and 660 km, respectively, π_1 and π_2 are the dimensionless excess pressures; Φ_1 and Φ_2 are the phase functions describing the relative fraction of the heavier phase, respectively, and varying between 0 and 1 as a function of depth and temperature. The Rayleigh (Ra), Laplace (La), and modified dissipation (Di^*) dimensionless numbers are defined as $Ra = \alpha g \rho^* T^* h^3 (\eta^* \kappa)^{-1}$, $La = \rho^* g h^3 (\eta^* \kappa)^{-1}$, and $Di^* = \eta^* \kappa (\rho^* c h^2 T^*)^{-1}$, respectively. The operator $\left(\mathbf{E} + \varsigma \nabla^2 \right)^{-1}$ is applied to the right-hand side of the Stokes equations (6.1) to smooth temperature jumps at the phase boundaries and to enhance the stability of our computations. According to the QRV method, the higher dissipation term, whose magnitude is controlled by the small parameter β, is introduced to regularize the heat balance Eq. (6.3). Length, temperature, and time are normalized by h, ΔT, and $h^2 \kappa^{-1}$, respectively. The physical parameters used in this study are listed in Table 6.1.

Table 6.1 Parameters of the numerical model

Parameter	Symbol	Value
Dimensionless density jump at the 410-km phase boundary	a_1	0.05
Dimensionless density jump at the 660-km phase boundary	a_2	0.09
Thermal conductivity	c	1250 W m^{-1} K^{-1}
Activation energy	E_a	3×10^5 J mol^{-1}
Acceleration due to gravity	g	9.8 m s^{-2}
Depth	h	800 km
Length (in x-direction)	l_1	4000 km
Length (in y-direction)	l_2	4000 km
Universal gas constant	R	8.3144 J mol^{-1} K^{-1}
Difference between the temperatures at the lower (T_l) and upper (T_u) model boundaries	T^*	1594 K
Dimensionless temperature at the upper model boundary	T_u	$290/T^*$
Dimensionless temperature at the lower model boundary	T_l	$1884/T^*$
Dimensionless temperature at the 410-km phase boundary	T_1	$1790/T^*$
Dimensionless temperature at the 660-km phase boundary	T_2	$1891/T^*$
Activation volume	V_a	4×10^{-6} m^3 mol^{-1}
Dimensionless width of the 410-km phase transition	w_1	10 km/h
Dimensionless width of the 660-km phase transition	w_2	10 km/h
Dimensionless depth of the 410-km phase boundary	z_1	390 km/h
Dimensionless depth of the 660-km phase boundary	z_2	140 km/h
Thermal expansivity	α	3×10^{-5} K^{-1}
QRV regularization parameter	β	0.00001
Dimensionless Clapeyron (pressure-temperature) slope at the 410-km phase boundary	$\overline{\gamma}_1$	4×10^6 Pa K$^{-1} \times T^*(\rho^* gh)^{-1}$
Dimensionless Clapeyron slope at the 660-km phase boundary	$\overline{\gamma}_2$	-2×10^6 Pa K$^{-1} \times T^*(\rho^* gh)^{-1}$
Reference viscosity	η^*	10^{21} Pa s
Thermal diffusivity	κ	10^{-6} m^{2}s^{-1}
Reference density	ρ^*	3400 kg m^{-3}
Phase regularization parameter	ς	0.0001

After Ismail-Zadeh et al. (2013)

6.3   Input Data: Seismic Temperature Model

The present temperature model beneath the Japanese islands is developed by using the high-resolution seismic tomography (P-wave velocity anomalies) for the region (Fukao et al. 2001; Obayashi et al. 2006, 2009). The seismic tomographic model consists of 16 horizontal layers of different thickness (12 km up to 88 km) starting from the depth of 12 km and extending down to the depth of 800 km. Each layer of 4000×4000 km^2 is subdivided horizontally into 80×80 km^2 blocks. To restrict numerical errors in our data assimilation, the velocity anomaly data are smoothed between the layers using a spline interpolation. The temperature anomalies are inferred from the seismic wave anomalies using a non-linear inversion method (Ismail-Zadeh et al. 2005) and considering the effects of mantle composition, anelasticity, and partial melting on seismic velocities (Karato and Wu 1993). The temperature anomalies are then added to the vertical temperature profile (the background temperature) to obtain the present temperature model. At shallow depth (down to 110 km), the background temperature is the solution of the cooling half-space model (Schubert et al. 2001) with 48 million years after the start of cooling. This temperature field gives a heat flow of 70 mW m^{-2}, which is close to the mantle heat flux. At deeper level, the background temperature is presented by the adiabatic temperature distribution (Katsura et al. 2004) with the potential temperature of 1330 °C. The crustal temperature inferred directly from the seismic velocity anomalies is relatively high, and the resulting heat flow is unrealistically elevated. Thus, the crustal temperature T_{cr} is calculated by estimating the geothermal gradient $\delta T_m / \delta x_3$ from measured regional surface heat flow (Wang et al. 1995; Yamano et al. 2003): $T_{cr}(x_1, x_2, x_3) = T_s + x_3 \cdot \delta T_m / \delta x_3 (x_1, x_2)$, where T_s is the surface temperature.

The present thermal state of the back-arc region so obtained is characterized by shallow hot anomalies reflecting the remnants of the back-arc spreading (Jolivet et al. 1994) and deep cold anomalies related to the stagnation of the lithospheric slabs (Fukao et al. 2001) (Fig. 6.2, see panels entitled "Present"). Meanwhile the state of the mantle beneath the Pacific plate is characterized by the shallow cold anomalies reflecting the existence of the old oceanic Pacific plate and the deep broad hot anomaly of unknown origin. This temperature model is used as the initial condition for restoration models.

6.4 Boundary Conditions

Although the mantle dynamics is coupled to the lithosphere dynamics, the coupling can be weak or strong depending on the viscosity contrast between the lithosphere and the underlying mantle (Doglioni et al. 2011). To explore how plate kinematics is related to the dynamics of the mantle, kinematic conditions are assumed at the upper model boundary, where the direction and rate of plate motion are prescribed. The plate motion velocity is determined from the Actual Plate Kinematic and Deformation Model (APKIM2005) derived from geodetic data (Drewes 2009) and from the PB2002 model for the Philippine Sea and Okinawa Plates (Bird 2003). The reference frame for the velocities is the International Terrestrial Reference Frame, which is realized by a set of stations on the Earth's surface. The reference for the velocities is defined with respect to Earth rotation, that is, there is no net rotation of

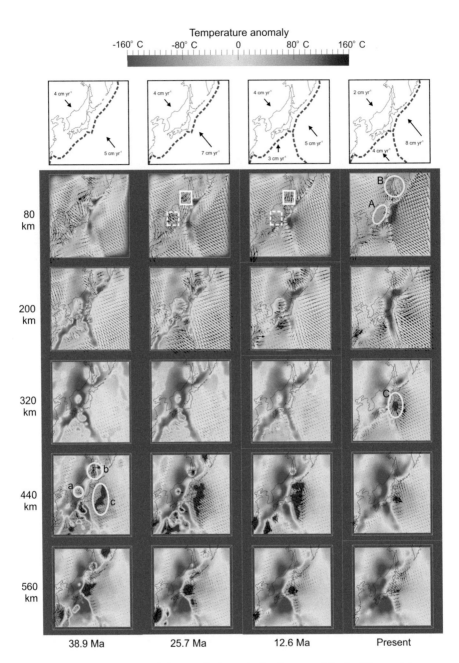

Fig. 6.2 Thermal evolution of the mantle beneath the Japanese Islands. The model covers 4000 km (horizontal) × 4000 km (horizontal) × 800 km (depth). The *top panel* presents the plate motion prescribed in the model. Horizontal cross-sections present temperature anomalies and the projection of mantle flow, which are obtained by the assimilation of the present temperature to the Middle Eocene times. *Yellow closed curves* indicate the hot mantle. *White boxes (bold and dashed)* show the places of mantle flow changes are likely to be associated with the rotation of north-eastern and southwestern parts of the Japan Sea in the Miocene times. See the text for characters *A, B, C, a, b,* and *c* (After Ismail-Zadeh et al. 2013)

the frame with respect to the rotating Earth (Drewes et al. 2006). The velocities are determined in APKIM2005 such a way to minimise a root mean square sum of all velocities over the Earth surface. The recent MORVEL model (Argus et al. 2011) is similar to APKIM2005, although the velocities in the MORVEL model are taken from different sources and have not the same reference frame.

Plate motions in the geological past are required to determine past mantle dynamics processes. The model incorporates the history of plate motions as the boundary conditions changing with time. Based on paleogeographic reconstructions of the Philippine Sea (Seno and Maruyama 1984) and Japanese Islands (Maruyama et al. 1997), Philippine Sea plate motion from paleomagnetic studies (Yamazaki et al. 2010), relative motion of the Pacific plate (Northrup et al. 1995), and Cenozoic plate tectonic evolution of the south-eastern Asia (Hall 2002), the velocity field is constructed to define the plate motion in the model for the past 40 million years, meanwhile recognizing the uncertainties in directions and magnitudes of the regional plate motion in the mentioned studies. To simplify the model, the present boundaries between the Okhotsk and Eurasian plates and between the North American and Pacific plates are omitted. Although change in the plate motion is a gradual process, for the sake of simplicity it is assumed that the plates moved with a constant velocity $W(t, x_1, x_2, x_3 = 0)$ within four time intervals in the past: from 0 to 10 million years ago (0–10 Ma), 10–20 Ma, 20–30 Ma, and 30–40 Ma (Fig. 6.2; see the upper panel).

Therefore, the velocity W and temperature $T = T_u$ are prescribed at the upper surface of the model boundary. Due to complexities and significant uncertainties in trench migration (oceanward versus trenchward) patterns observed on the wide Pacific and Philippine Sea subduction zone (Schellart et al. 2008), the trench migration is not included in the model by Ismail-Zadeh et al. (2013). Also a weak (artificial) zone at the plate interfaces is not introduced (as it is used in forward modelling to reduce stresses at the interfaces and to promote subduction). Such a weak zone is partly accommodated in backward modelling, because in the backward (time-reverse) modelling the hotter material from below moves upward toward the divergent plate interfaces reducing the viscosity there. In numerical models of mantle dynamics, when a surface velocity is prescribed, the flow is driven by a combination of the velocity and internal buoyancy of the fluid. The choice of the prescribed velocities affects the flow velocities in the uppermost mantle. To analyse the influence of the prescribed velocity, the velocities have been varied by changing their magnitude (within about 3 %) and direction (within about ±2 degrees). The small changes in the imposed velocities are found not to change significantly the mantle flow and hence the pattern of upwellings.

The velocity $u = 0$ (no-slip) and fixed temperature $T = T_l$ are prescribed at the lower surface of the model boundary. To allow for the flow to pass through the lateral boundaries, the conditions $\partial u / \partial n = 0$ and $\partial T / \partial n = 0$ are introduced at the lateral sides of the boundary. At the model boundary $\partial P / \partial n = 0$ is assumed. Several other conditions at the lateral sides and at the bottom of the model domain have been tested by Ismail-Zadeh et al. (2013) to analyse how the boundary conditions influence the restoration results.

6.5 Rheological Model

Mantle rheology is rather complex and depends on temperature and pressure (Gordon 1967), stresses (Hirth and Kohlstedt 2003), water content (Hirth and Kohlstedt 1996), grain size (Karato et al. 2001), and composition (Lee and Chen 2007). Slow deformation of minerals in the upper mantle under the influence of stresses and temperature is governed by diffusion and dislocation creeps (Hirth and Kohlstedt 2003). Diffusion creep dominates deformation at cooler temperatures and larger grain sizes, whereas dislocation creep dominates deformation at higher strain rates and is not grain-size dependent (Billen 2008). Diffusion-controlled creep is characterized by the Newtonian rheology, meanwhile dislocation creep by a non-Newtonian rheology (e.g., a power-law relationship between stress and strain rate). If in the upper mantle diffusion and dislocation creep mechanisms act together to accommodate deformation, the lower mantle is likely to deform by diffusion creep as seismic anisotropy studies detect less seismic anisotropy in the deep mantle (Savage 1999).

Effects of a non-Newtonian viscosity on steady and time-dependent convection have been studied intensively using 2-D numerical models (e.g., Christensen 1984; Christensen and Yuen 1989; Malevsky and Yuen 1992; Billen and Hirth 2005) and 3-D numerical models (e.g., Stadler et al. 2010). For a stationary convection of a temperature- and pressure-dependent viscosity fluid it was shown that the use of the Newtonian viscosity with activation enthalpy one-third to half of the experimentally determined value for olivine could mimic the dynamics of the convection with a strongly non-Newtonian power-law viscosity (Christensen 1984). Although a non-stationary thermal convection introduces changes in the pattern of viscous flow, the qualitative behaviour of the flow is similar, and the difference comes in the way of time-dependence.

In the numerical modelling Ismail-Zadeh et al. (2013) assumed that the Earth's mantle behaves as a temperature- and pressure-dependent (Arrhenius-type) Newtonian fluid $\eta\left(T\left(\mathbf{x}\right), x_3\right) = \eta_0 \exp\left[\frac{E_a + \rho_* g x_3 V_a}{RT}\right]$, where η_0 is determined so that it will give 2.905×10^{20} Pa s at the depth of 290 km and temperature of 1698 K; the activation energy is $E_a = 3 \times 10^5$ J mol^{-1}, and the activation volume of $V_a = 4 \times 10^{-6}$ m^3 mol^{-1}. Other parameters of the rheological law are listed in Table 6.1. The upper limit of the viscosity is set to be $\sim 10^{22}$ Pa s, which results in the viscosity increase from the upper to the lower mantle by about two orders of magnitude.

According to experimental studies of olivine deformation, the values of activation energy and activation volume in a dry upper mantle are 3×10^5 J mol^{-1} and 6×10^{-6} m^3 mol^{-1} for diffusion creep, and 5.4×10^5 J mol^{-1} and 2×10^{-5} m^3 mol^{-1} for dislocation creep, respectively (Karato and Wu 1993). Measurements of the activation volume for creep of dry olivine at pressures of 2.7–4.9 GPa and temperatures near 1473° K showed a wide range of the values $9.5 \pm 7 \times 10^{-6}$ m^3 mol^{-1} (Durham et al. 2009). For the prescribed model values of the activation energy and the activation volume, the model activation enthalpy is less than half of the experimentally determined values, and therefore, the Newtonian fluid with the chosen model values can approximate to some extent the mechanical behaviour of the upper mantle. The inclusion of "realistic" rheology of the upper mantle in the model would be preferable. The numerical experiments using non-Newtonian power-law rheology showed that, although there exists some differences between two rheological models (Newtonian linear and non-linear rheologies), the overall mantle dynamics is rather similar (unpublished work by S. Honda).

6.6 Numerical Approach

The problem (6.1), (6.2) and (6.3) with the prescribed boundary and input conditions have been solved by the finite-volume method (Ismail-Zadeh et al. 2013) using open source computational fluid dynamics software package OpenFoam (http://www.openfoam.com). As $200 \times 200 \times 190$ finite volumes (rectangular hexahedrons) are used, a horizontal resolution of the model is 20 km $\times$ 20 km. The model domain is divided into five horizontal layers: Layer 1 (from the surface to the depth of 400 km), layer 2 (400–420 km), layer 3 (420–650 km), layer 4 (650–670 km), and layer 5 (670–800 km). Within each layer 60, 40, 35, 40, and 15 grid points are used. Therefore, a vertical resolution of the model varies from 0.5 to 8.67 km.

Velocity $\mathbf{u}$ and pressure P are found from the Eqs. (6.1) and (6.2) using the SIMPLE method (Patankar and Spalding 1972). The regularized heat balance Eq. (6.3) is approximated by the Euler method using the implicit approximation of the advective term and the explicit approximation of the conductive term:

$$(\mathbf{E} + \beta \mathbf{D})^2 \frac{T^{n+1} - T^n}{dt} + \mathbf{C} T^{n+1} - \mathbf{D} T^n + f(\mathbf{u}, T^n) = 0,$$

where the discrete operators $\mathbf{C} = -\mathbf{C}^T$ and $\mathbf{D} = \mathbf{D}^T$ approximate the advective and conductive terms, respectively. To solve the numerical scheme the splitting method is used (Samarskii and Vabischevich 1995) by introducing the convection/anti-diffusion and regularization parts as

$$(\mathbf{E} + dt\mathbf{C}) T^{n+1/2} = (\mathbf{E} + dt\mathbf{D}) T^n - dt f(\mathbf{u}, T^n), \qquad (6.4)$$

$$(\mathbf{E} + \beta \mathbf{D})^2 T^{n+1} = T^{n+1/2}. \qquad (6.5)$$

The system of the discrete equations (6.4) is solved by the BiConjugate Gradient method (Van der Vorst 1992) using the incomplete LU-factorization as a preconditioner (Saad 1996). The system (6.5) is solved by the conjugate gradient method (e.g., Ismail-Zadeh and Tackley 2010).

6.7 Model Results

In backward sense, the high-temperature patchy anomaly beneath the back-arc Japan Sea basin splits into two prominent anomalies showing two small-scale upwellings beneath the southwestern and northern part of the Japan Sea (Fig. 6.2). The present hot anomalies in the back-arc region marked by A and B move down to the spots marked by a and b at 38.9 Ma (Fig. 6.2). Meanwhile the broad hot anomaly under the Pacific plate moves slowly down westward as depicted by C at present and c at 38.9 Myr in Fig. 6.2. The hot anomalies (spot b) and (spot c) tend to merge at 38.9 Ma and below 560 km. The model shows the link between spot b in the back-arc region and spot c in the sub-slab mantle at depths of about 440–560 km in the Middle to Late Eocene time. The upwellings a and b are likely to be generated in the sub-slab hot mantle (Figs. 6.2 and 6.3) and penetrated through breaches/tears of the subducting Pacific plate into the mantle wedge toward spots A and B. Hence, Ismail-Zadeh et al. (2013) proposed that the present hot anomalies in the back-arc and sub-slab mantle had a single origin located in the sub-lithospheric mantle.

The shallow mantle velocity field in the back arc, which potentially affects the surface tectonics there, shows rather complex pattern with time compared to that in other area. The south-westward uppermost mantle flow (the boxed area at depth 80 km of Fig. 6.2) beneath the north-eastern (NE) part of the Japan Sea region in the Middle Eocene – Oligocene times, 38.9–25.7 Ma, changed to the south-south-eastward flow in the Miocene times, 12.6 Ma (Fig. 6.2). This flow change could contribute to counterclockwise rotation of the NE Japan Islands, which resulted in the back-arc basin opening (e.g., Otofuji et al. 1985; Tatsumi et al. 1989). In the southwestern (SW) Japan, clockwise rotation was likely to be instantaneous at about 15 Ma (Otofuji et al. 1985), which led to opening of this part of the Japan Sea. Although the model cannot explicitly predict the changes in the mantle flow in the SW Japan Sea region, the flow pattern in the uppermost mantle (the dash boxed area

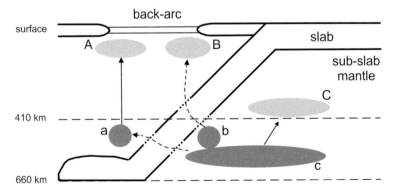

Fig. 6.3 Schematic representation of the thermal evolution beneath the Japanese islands and their surroundings (see the text) (After Ismail-Zadeh et al. 2013)

at 80 km depth of Fig. 6.2) indicates that the north-westward flow established in Oligocene becomes much weaker in Miocene.

An analysis of the onshore and offshore borehole data from South Sakhalin, the Oga Peninsula, and Dolgorae shows that the regional tectonic subsidence commenced in the Early Oligocene and continued until the Pliocene with a few breaks for a tectonic uplift (Fig. 6.4; Ingle 1992). The rapid subsidence of South Sakhalin (in the Early Miocene, some 23–22 Myr ago) was followed by the slow subsidence and uplift until about 16 Myr ago, which was interrupted by the second phase of rapid subsidence in the Middle Miocene (15 Myr ago) followed again by slow subsidence and uplift. The alternation of subsidence and uplift phases in the basin is indicative of rifting style changes from rollback-induced passive rifting/extension to upwelling-induced (upwelling B, Figs. 6.3 and 6.4) active rifting/extension (Huismans et al. 2001). Thus the small-scale upwelling beneath the northern part of the Japan Sea (predicted by the data assimilation modelling) could contribute to the rifting and back arc basin opening. The evolution of the southwestern Japan Sea recorded in the sediments (drilled by the offshore Dolgarae-1 borehole) shows a persistent subsidence of the basin until about 11 Myr ago followed by rapid uplift. This uplift is likely to be associated with the small-scale upwelling A (Figs. 6.3 and 6.4). Meanwhile preceded by a long phase (about

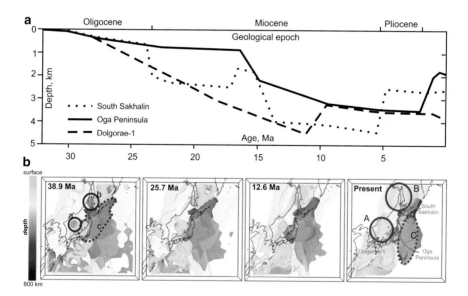

Fig. 6.4 Subsidence history of the Japan Sea: tectonic subsidence curves (Ingle 1992) for the southern Sakhalin section representing the northern margin of the Japan Sea (*dotted curve*), the Oga Peninsula section representing the inner arc area of north-western Honshu (*solid curve*), and Dolgorae-1 well representing the southern Tsushima Basin and the southern margin of the Japan Sea (*dashed curve*). The location of the wells is marked in (*b*). (*b*) Three-dimensional view of snapshots of the iso-surfaces of positive (5 %) temperature anomalies; colours mark the depth (After Ismail-Zadeh et al. 2013)

15 Myr) of a slow subsidence, the relatively rapid subsidence of the Oga Peninsula (which is located in-between two upwellings A and B) started in the Middle Miocene (about 16 Myr ago) followed by slow subsidence for about 13 Myr and then rapid uplift in Quaternary (about 2 Myr ago).

If the hypothesis, that the hot mantle upwellings under the Japan Sea originated beneath the Pacific plate and penetrated through the subducting lithosphere contributing to the back-arc spreading, is valid, a signature of such a phenomenon may be found in magmatic rock samples. The change in the mantle source of magmatic rocks related to the opening of the Japan Sea from enriched to depleted with time (Tatsumi et al. 1989) suggests a possibility for a physical replacement of the enriched subcontinental upper mantle with the depleted asthenospheric mantle by an injection during back-arc development. The sample showing the mixture of asthenospheric and lithospheric slab material is found (Hanyu et al. 2006) near the triple junction of the Pacific, Okhotsk and Philippine Sea plates (shown by star in Fig. 6.1). The age of this sample is about 23 Ma and corresponds to early stages of the Japan Sea opening. These observations can be explained by our hypothesis of hot upwellings coming from the Pacific side.

Presently, there is no consensus on the origin of hot anomaly under the Pacific plate. It can be a relic of the hot plume originated at the thermal boundary layers, although its closeness to the cold subduction zone seems to reject the nearby origin of hot upwellings. It can be a relic of large plume head conveyed horizontally by the plate movement or the hot thermal anomaly close to the sinking plume typical to the internally heated convection. The previous study shows that the latter scenario is more likely (Morishige et al. 2010). There is also a suggestion that the high temperature anomaly adjacent to the cold plume in internally heated convection may even reverse the overall direction of flow implying that the dynamical effect of such a high temperature anomaly is fairly significant.

It is known that the opening of the Japan Sea is not a simple homogeneous and instantaneous spreading, but has temporal and spatial variations and shows several stages of deformation (Jolivet et al. 1994). Inhomogeneous spreading is consistent with the patchy character of hot materials as evident in our results. Non-instantaneous deformation may imply that the hot materials have penetrated through, or affected, the overlying subducting Pacific lithosphere several times.

The slab break-off, detachment and tear have been intensively investigated for the last decades. A lithospheric slab tear has been proposed in the Indonesian arc (Widiyantoro and van der Hilst 1996) and in the Mediterranean region (Wortel and Spakman 2000). A horizontal tear in the subducting slab has been recognized in the Izu-Bonin-Mariana arc (e.g., Miller et al. 2005). There are numerical studies related to this phenomenon (Durez et al. 2011), but they usually do not assume the existence of hot anomalies under the subducting lithosphere. Morishige et al. (2010) showed a possibility of slab break-off by hot materials in the sub-lithospheric mantle. However, the model results by Ismail-Zadeh et al. (2013) do not show the slab break-off but the penetration via breaches/tears in the subducting lithosphere, which is probably more likely than the total break-off.

The geometry of the thermal structures in the mantle changes with time due to heat advection, which deforms the structures, and heat conduction, which smoothes the complex shapes of the structures. This creates difficulties in understanding the evolution of the mantle structures in the past. A quantitative assimilation of the present mantle temperature and flow into the geological past provides a tool for restoration of thermal shapes of prominent structures in the past from their diffusive shapes at present. Using the present seismic tomography data as well as some other geological, geophysical and geodetic data and constraints, Ismail-Zadeh et al. (2013) reconstructed prominent features of hot upwelling beneath the Japan Sea and Japanese islands using the QRV method.

There are several principal qualitative models proposed to explain back-arc spreading: mantle diapirism (Karig 1971), convection induced by the subducting slab (Sleep and Toksoz 1971), movements of adjacent plates (Uyeda and Kanamori 1979), and the injection of hot asthenospheric material (Tatsumi et al. 1989). The model by Ismail-Zadeh et al. (2013) (see Fig. 6.3) as the latter one supports the hypothesis of an asthenospheric injection. Meanwhile the major difference between the two models is related to where the hot material came from. While the model by Tatsumi et al. (1989) assumes that it may have come from the back-arc side, the results of dynamic restoration of the thermal state of the mantle in the past by Ismail-Zadeh et al. (2013) show that it came from the opposite side, that is, from the mantle under the Pacific plate. High-resolution experiments on seismic wave attenuation, improved knowledge of crustal and mantle mineral composition, and enhanced paleo-reconstruction models of plate movements in the region will assist to refine the present model of the mantle thermal evolutions beneath the western Pacific region.

6.8 Data Uncertainties

Data assimilation can be influenced primarily by uncertainties of a present temperature model (initial condition) and uncertainties in boundary conditions. Uncertainties in the temperature model used in this study come from seismic tomography models, mantle composition, seismic attenuation models, and poor knowledge of the presence of water or melt content at mantle depths. Seismic tomography models may introduce some uncertainties in data assimilation due to a resolution power. Ismail-Zadeh et al. (2013) compared the images related to different seismic tomography models and showed that the resolution power of the models is almost similar.

The temperature at the lower boundary of the model domain used in the numerical modelling is an approximation to the real temperature, which is unknown and may change over time at this boundary. The conditions at lateral boundaries are prescribed to satisfy certain properties of the model (e.g., conservation of mass, conservation of momentum), meanwhile the true conditions at the boundaries are unknown and hence contribute to uncertainties of the model. The velocities prescribed at the upper boundary of the model domain comes from recent geodetic observations and paleogeographic/paleomagnetic reconstructions of plate motion, and hence the uncertainties in the prescribed velocity field are associated with an accuracy of geodetic measurement, viscosity contrast between the lithosphere and the underlying mantle, and uncertainties in paleo-reconstructions. In numerical modelling sensitivity analysis assists in understanding the stability of model solutions to small perturbations in input variables or parameters.

Ismail-Zadeh et al. (2013) performed a sensitivity analysis to understand how stable is the numerical solution to small perturbations of input data (the present temperature). The model of the present temperature has been perturbed randomly by 0.5–1 % and then assimilated to the past to find the initial temperature. A misfit between the initial temperatures related to the perturbed and unperturbed present temperature is about 3–5 %, which proves the stability of the solution.

For a given temperature, mantle dynamics depends on various factors including rheology (Billen 2008), phase changes (Liu et al. 1991; Honda et al. 1993a, b), and boundary conditions. Ismail-Zadeh et al. (2013) conducted a search over the ranges of uncertain parameters in the temperature- and pressure-dependent viscosity (activation energy and activation volume) to achieve 'plate-like' behaviour of the colder material. Also they tested the influence of phase changes, model depth variations and boundary conditions on the model results. In numerical models of mantle dynamics the choice of boundary conditions and the size of the model domain influence the pattern of flow and slab dynamics. If the depth of the model domain is significantly smaller than the horizontal dimensions of the domain, the thermo-convective flow in the model with a low viscosity upper mantle and higher viscosity lower mantle will generate the return flow focused in the upper mantle. Increasing the model domain's depth removes the artificial lateral return flow in the upper mantle. The sensitivity analysis related to the presence of phase transformations and to changes in boundary conditions show that the model is robust, and the principal results of the model do not change (Ismail-Zadeh et al. 2013).

References

Argus DF, Gordon RG, DeMets C (2011) Geologically current motion of 56 plates relative to the no-net-rotation reference frame. Geochem Geophys Geosyst 12:Q11001. doi:10.1029/2011GC003751

Billen MI (2008) Modeling the dynamics of subducting slabs. Annu Rev Earth Planet Sci 36:325–356

Billen MI, Hirth G (2005) Newtonian versus non-Newtonian upper mantle viscosity: implications for subduction initiation. Geophys Res Lett 32:L19304. doi:10.1029/2005GL023457

Bird P (2003) An updated digital model of plate boundaries. Geochem Geophys Geosyst 4:1027. doi:10.1029/2001GC000252

Christensen U (1984) Convection with pressure and temperature-dependent non-Newtonian rheology. Geophys J Roy Astron Soc 77:343–384

Christensen UR, Yuen DA (1985) Layered convection induced by phase transitions. J Geophys Res 90:10291–10300

Christensen U, Yuen D (1989) Time-dependent convection with non-Newtonian viscosity. J Geophys Res 94:814–820

Doglioni C, Ismail-Zadeh A, Panza G, Riguzzi F (2011) Lithosphere-asthenosphere viscosity contrast and decoupling. Phys Earth Planet Inter 189:1–8

Drewes H (2009) The actual plate kinematic and crustal deformation model APKIM2005 as basis for a non-rotating ITRF. In: Drewes H (ed) Geodetic reference frames, IAG symposia series 134. Springer, Berlin, pp 95–99

Drewes H, Angermann D, Gerstl M, Krügel M, Meisel B, Seemüller W (2006) Analysis and refined computations of the international terrestrial reference frame. In: Flury J, Rummel R, Reigber C, Rothacher M, Boedecker G, Schreiber U (eds) Observation of the earth system from space. Springer, Heidelberg, pp 343–356

Durez T, Gerya TV, May DA (2011) Numerical modelling of spontaneous slab breakoff and subsequent topographic response. Tectonophysics 502:244–256

Durham WB, Mei S, Kohlstedt DL, Wang L, Dixon NA (2009) New measurements of activation volume in olivine under anhydrous conditions. Phys Earth Planet Inter 172:67–73

Fukao Y, Widiyantoro S, Obayashi M (2001) Stagnant slabs in the upper and lower mantle transition region. Rev Geophys 39:291–323

Furumura T, Kennett BLN (2005) Subduction zone guided waves and the heterogeneity structure of the subducted plate – intensity anomalies in northern Japan. J Geophys Res 110:B10302. doi:10.1029/2004JB003486

Gordon RB (1967) Thermally activated processes in the Earth: Creep and seismic attenuation. Geophys J Roy Astron Soc 14:33–43

Hall R (2002) Cenozoic geological and plate tectonic evolution of SE Asia and the SW Pacific: computer-based reconstructions, model and animations. J Asian Earth Sci 20:353–431

Hanyu T, Tatsumi Y, Nakai S, Chang Q, Miyazaki T, Sato K, Tani K, Shibata T, Yoshida T (2006) Contribution of slab melting and slab dehydration to magmatism in the NE Japan arc for the last 25 Myr: constraints from geochemistry. Geochem Geophys Geosyst 7:Q08002. doi:10.1029/2005GC001220

Hirth G, Kohlstedt DL (1996) Water in the oceanic upper mantle: implications for rheology, melt extraction and the evolution of the lithosphere. Earth Planet Sci Lett 144:93–108

Hirth G, Kohlstedt DL (2003) Rheology of the upper mantle and the mantle wedge: a view from the experimentalists. In: Eiler J (ed) Inside the subduction factory. Geophysical monograph. 138. American Geophysical Union, Washington, DC, pp 83–105

Honda S, Balachandar S, Yuen DA, Reuteler D (1993a) Three-dimensional mantle dynamics with an endothermic phase transition. Geophys Res Lett 20:221–224

Honda S, Yuen DA, Balachandar S, Reuteler D (1993b) Three-dimensional instabilities of mantle convection with multiple phase transitions. Science 259:1308–1311

Huismans RS, Podladchikov YY, Cloetingh S (2001) Transition from passive to active rifting: relative importance of asthenospheric doming and passive extension of the lithosphere. J Geophys Res 106:11271–11291

Ingle CJ (1992) Subsidence of the Japan Sea: stratigraphic evidence from ODP sites and onshore sections. In: Tamaki K, Suyehiro K, Allan J, McWilliams et al (eds) Proceedings of the ocean drilling program scientific results, 127/128, Pt. 2. Ocean Drilling Program, College Station, pp 1197–1218

Ismail-Zadeh A, Tackley P (2010) Computational methods for geodynamics. Cambridge University Press, Cambridge

Ismail-Zadeh A, Mueller B, Schubert G (2005) Three-dimensional modeling of present-day tectonic stress beneath the earthquake-prone southeastern Carpathians based on integrated analysis of seismic, heat flow, and gravity observations. Phys Earth Planet Inter 149:81–98

Ismail-Zadeh A, Korotkii A, Schubert G, Tsepelev I (2007) Quasi-reversibility method for data assimilation in models of mantle dynamics. Geophys J Int 170:1381–1398

Ismail-Zadeh A, Honda S, Tsepelev I (2013) Linking mantle upwelling with the lithosphere descent and the Japan Sea evolution: a hypothesis. Sci Rep 3:1137. doi:10.1038/srep01137

Jolivet L, Tamaki K, Fournier M (1994) Japan Sea, opening history and mechanism: a synthesis. J Geophys Res 99:22232–22259

Karato S, Wu P (1993) Rheology of the upper mantle: a synthesis. Science 260:771–778

Karato S-I, Riedel MR, Yuen DA (2001) Rheological structure and deformation of subducted slabs in the mantle transition zone: implications for mantle circulation and deep earthquakes. Phys Earth Planet Inter 127:83–108

Karig DE (1971) Origin and development of marginal basins in the Western Pacific. J Geophys Res 76:2542–2561

Katsura T, Yamada H, Nishikawa O, Song M, Kubo A, Shinmei T, Yokoshi S, Aizawa Y, Yoshino T, Walter MJ, Ito E, Funakoshi K (2004) Olivine-wadsleyite transition in the system (Mg, Fe)$_2$SiO$_4$. J Geophys Res 109:B02209. doi:10.1029/2003JB002438

Lee CTA, Chen WP (2007) Possible density segregation of subducted oceanic lithosphere along a weak serpentinite layer and implications for compositional stratification of the Earth's mantle. Earth Planet Sci Lett 255:357–366

Liu M, Yuen DA, Zhao W, Honda S (1991) Development of diapiric structures in the upper mantle due to phase transitions. Science 252:1836–1839

Malevsky AV, Yuen DA (1992) Strongly chaotic non-Newtonian mantle convection. Geophys Astrophys Fluid Dyn 65:149–171

Maruyama S, Isozaki Y, Kimura G, Terabayashi M (1997) Paleogeographic maps of the Japanese islands: plate tectonic synthesis from 750 Ma to the present. Island Arc 6:121–142

Miller MS, Gorbatov A, Kennett BLN (2005) Heterogeneity within the subducting Pacific plate beneath the Izu-Bonin-Mariana arc: evidence from tomography using 3D ray-tracing inversion techniques. Earth Planet Sci Lett 235:331–342

Morishige M, Honda S, Yoshida M (2010) Possibility of hot anomaly in the sub-slab mantle as an origin of low seismic velocity anomaly under the subducting Pacific plate. Phys Earth Planet Inter 183:353–365

Northrup CJ, Royden LH, Burchfiel BC (1995) Motion of the Pacific plate relative to Eurasia and its potential relation to Cenozoic extension along the eastern margin of Eurasia. Geology 23:719–722

Obayashi M, Sugioka H, Yoshimitsu J, Fukao Y (2006) High temperature anomalies oceanward of subducting slabs at the 410-km discontinuity. Earth Planet Sci Lett 243:149–158

Obayashi M, Yoshimitsu J, Fukao Y (2009) Tearing of stagnant slab. Science 324:1173–1175

Otofuji Y, Matsuda T, Nohda S (1985) Opening mode of the Japan Sea inferred from the paleomagnetism of the Japan arc. Nature 317:603–604

Patankar SV, Spalding DB (1972) A calculation procedure for heat and mass transfer in three-dimensional parabolic flows. Int J Heat Mass Transf 15:1787–1806

Saad Y (1996) Iterative methods for sparse linear systems. PWS, Boston

Samarskii AA, Vabishchevich PN (1995) Computational heat transfer. Vol. 2. The finite difference methodology. Wiley, New York

Savage MK (1999) Seismic anisotropy and mantle deformation: what have we learned from shear wave splitting. Rev Geophys 374:65–106

Schellart WP, Stegman DR, Freeman J (2008) Global trench migration velocities and slab migration induced upper mantle volume fluxes: constraints to find an earth reference frame based on minimizing viscous dissipation. Earth Sci Rev 88:118–144

Schubert G, Turcotte DL, Olson P (2001) Mantle convection in the earth and planets. Cambridge University Press, Cambridge

Seno T, Maruyama S (1984) Paleogeographic reconstruction and origin of the Philippine Sea. Tectonophysics 102:53–84

Sleep N, Toksoz MN (1971) Evolution of marginal basins. Nature 233:548–550

Stadler G, Gurnis M, Burstedde C, Wilcox LC, Alisic L, Ghattas O (2010) The dynamics of plate tectonics and mantle flow: from local to global scales. Science 329:1033–1038

Tatsumi Y, Otofuji Y, Matsuda T, Nohda S (1989) Opening of the Japan Sea by asthenospheric injection. Tectonophysics 166:317–329

Uyeda S, Kanamori H (1979) Back-arc opening and the model of subduction. J Geophys Res 84:1049–1061

Van der Vorst HA (1992) BI-CGSTAB: a fast and smoothly converging variant of BI-CG for the solution of nonsymmetric linear systems. SIAM J Sci Stat Comput 13(2):631–644

Wang K, Hyndman RD, Yamano M (1995) Thermal regime of the Southwest Japan subduction zone: effects of age history of the subducting plate. Tectonophysics 248:53–69

Widiyantoro S, van der Hilst RD (1996) Structure and evolution of subducted lithosphere beneath the Sunda Arc. Science 271:1566–1570

Wortel MJR, Spakman W (2000) Subduction and slab detachment in the Mediterranean–Carpathian region. Science 290:1910–1917

Yamano M, Kinoshita M, Goto S, Matsubayashi O (2003) Extremely high heat flow anomaly in the middle part of the Nankai Trough. Phys Chem Earth 28:487–497

Yamazaki T, Takahashi M, Iryu Y, Sato T, Oda M, Takayanagi H, Chiyonobu S, Nishimura A, Nakazawa T, Ooka T (2010) Philippine Sea Plate motion since the Eocene estimated from paleomagnetism of seafloor drill cores and gravity cores. Earth Planets Space 62:495–502

Chapter 7
Comparison of Data Assimilation Methods

Abstract Following Ismail-Zadeh et al. (Geophys J Int 170:1381–1398, 2007), we compare in this chapters the backward advection (BAD), variational (VAR), and quasi-reversibility (QRV) methods in terms of solution stability, convergence, and accuracy, time interval for data assimilation, analytical and algorithmic works, and computer performance.

Keywords Backward advection • Variational method • Quasi-reversibility • Data assimilation

We have presented in this book three basic methods for data assimilation in geo-dynamic models and illustrated their applicability using several case studies. Each method has its advantages and a number of disadvantages. Table 7.1 summaries the differences in the methodology of data assimilation. The VAR data assimilation assumes that the direct and adjoint problems are constructed and solved iteratively forward in time. The structure of the adjoint problem is identical to the structure of the original problem, which considerably simplifies the numerical implementation. However, the VAR method imposes some requirements for the mathematical model (i.e. a derivation of the adjoint problem). Moreover, for an efficient numerical implementation of the VAR method, the error level of the computations must be adjusted to the parameters of the algorithm, and this complicates computations.

The QRV method allows employing sophisticated mathematical models (because it does not require derivation of an adjoint problem as in the VAR data assimilation) and hence expands the scope for applications in geodynamics (e.g. thermo-chemical convection, phase transformations in the mantle). It does not require that the desired accuracy of computations be directly related to the parameters of the numerical algorithm. However, the regularizing operators usually used in the QRV method enhance the order of the system of differential equations to be solved.

The BAD method does not require any additional work (neither analytical nor computational). The major difference between the BAD method and two other methods (VAR and QRV methods) is that the BAD method is by design expected to work (and hence can be used) *only* in advection-dominated heat flow. In the regions of high temperature/low mantle viscosity, where heat is transferred mainly by convective flow, the use of the BAD method is justified, and the results of

A. Ismail-Zadeh et al., *Data-Driven Numerical Modelling in Geodynamics: Methods and Applications*, SpringerBriefs in Earth Sciences,
DOI 10.1007/978-3-319-27801-8_7

Table 7.1 Comparison of data assimilation methods in geodynamic models

Method	QRV method	VAR method	BAD method
Method	Solving the regularized backward heat problem with respect to parameter β	Iterative sequential solving of the direct and adjoint heat problems	Solving of heat advection equation backward in time
Solution's stability	Stable for parameter β to numerical errors (see text; also in[1]) and conditionally stable for parameter β to arbitrarily assigned initial conditions (numerically[2])	Conditionally stable to numerical errors depending on the number of iterations (theoretically[3]) and unstable to arbitrarily assigned initial conditions (numerically[4])	Stable theoretically and numerically
Solution's convergence	Numerical solution to the regularized backward heat problem converges to the solution of the backward heat problem in the special class of admissible solutions[5]	Numerical solution converges to the exact solution in the Hilbert space[6]	Not applied
Solution's accuracy[7]	Acceptable accuracy for both synthetic and geophysical data	High accuracy for synthetic data	Low accuracy for both synthetic and geophysical data in conduction-dominated mantle flow
Time interval for data assimilation[8]	Limited by the characteristic thermal diffusion time	Limited by the characteristic thermal diffusion time and the accuracy of the numerical solution	No specific time limitation; depends on mantle flow intensity
Analytical work	Choice of the regularizing operator	Derivation of the adjoint problem	No additional analytical work
Algorithmic work	New solver for the regularized equation should be developed	No new solver should be developed	Solver for the advection equation is to be used

After Ismail-Zadeh et al. (2007)
[1]Lattes and Lions 1969; [2]see Fig. 5.3 and relevant text; [3]Ismail-Zadeh et al. 2004; [4]Ismail-Zadeh et al. 2006; [5]Tikhonov and Arsenin 1977; [6]Tikhonov and Samarskii 1990; [7]see Table 7.2; [8]see text for details

Table 7.2 Quality of the numerical results obtained by different methods

Quality	Synthetic data		Geo-data	
	Advection-dominated regime	*Diffusion-dominated region*	*Advection-dominated regime*	*Diffusion-dominated region*
Good	VAR	VAR	–	–
Satisfactory	QRV, BAD	QRV	QRV, BAD	QRV
Poor	–	BAD	–	BAD

After Ismail-Zadeh et al. (2007)

numerical reconstructions can be considered to be satisfactory. Otherwise, in the regions of conduction-dominated heat flow (due to either high mantle viscosity or high conductivity of mantle rocks), the use of the BAD method cannot even guarantee any similarity of reconstructed structures. If mantle structures are diffused significantly, the remaining features of the structures can be only backward advected with the flow.

The comparison between the data assimilation methods is summarized in Table 7.2 in terms of a quality of numerical results. The quality of the results is defined here as a relative (not absolute) measure of their accuracy. The results are good, satisfactory, or poor compared to other methods for data assimilation considered in this study. The numerical results of the reconstructions for both synthetic and geophysical case studies show the comparison quantitatively.

The time interval for the VAR data assimilation depends strongly on smoothness of the input data and the solution. The time interval for the BAD data assimilation depends on the intensity of mantle convection: it is short for conduction-dominated heat transfer and becomes longer for advection-dominated heat flow. In the absence of thermal diffusion the backwards advection of a low-density fluid in the gravity field will finally yield a uniformly stratified, inverted density structure, where the low-density fluid overlain by a dense fluid spreads across the lower boundary of the model domain to form a horizontal layer. Once the layer is formed, information about the evolution of the low-density fluid will be lost, and hence any forward modelling will be useless, because no information on initial conditions will be available (Ismail-Zadeh et al. 2001).

The QRV method can provide stable results within the characteristic thermal diffusion time interval. However, the length of the time interval for QRV data assimilation depends on several factors. Let us explain this by the example of heat conduction Eq. (5.1). Assume that the solution to the backward heat conduction equation with the boundary conditions (5.2) and the initial condition $T(t = t^*, x) = T^*(x)$ satisfies the inequality $\left\| \partial^4 T / \partial x^4 \right\| \leq L_d$ at any time t. This strong additional requirement can be considered as the requirement of sufficient smoothness of the solution and initial data. Considering the regularized backward heat conduction Eq. (5.5) with the boundary conditions (5.6)–(5.7) and the input temperature $T_\beta(t = t^*, x) = T_\beta^*(x)$ and assuming that $\left\| T_\beta^* - T^* \right\| \leq \delta$, Samarskii and Vabishchevich (2007) estimated the temperature misfit between the solution

Table 7.3 Performance of data assimilation methods

Method	CPU time (circa, in s)		Total
	Solving the Stokes problem using $50 \times 50 \times 50$ finite elements	Solving the backward heat problem using $148 \times 148 \times 148$ finite difference mesh	
BAD	180	2.5	182.5
QRV	100–180	3	103–183
VAR	360	$1.5\,n$	$360 + 1.5\,n$

After Ismail-Zadeh et al. (2007)

$T(t, x)$ to the backward heat conduction problem and the solution $T_\beta(t, x)$ to the regularized backward heat conduction equation:

$$\left\| T\left(t, x\right) - T_\beta\left(t, x\right) \right\| \leq \tilde{C}\delta \exp\left[\beta^{-1/2}\left(t^* - t\right)\right] + \beta L_d t, \ 0 \leq t \leq t^*,$$

where constant $\tilde{C}$ is determined from the a priori known parameters of the backward heat conduction problem. For the given regularization parameter β, errors in the input data δ, and smoothness parameter L_d, it is possible to evaluate the time interval $0 \leq t \leq t^*$ of data assimilation for which the temperature misfit would not exceed a prescribed value.

Computer performance of the data assimilation methods can be estimated by a comparison of CPU times for solving the inverse problem of thermal convection. Table 7.3 lists the CPU times required to perform one time-step computations on 16 processors. The CPU time for the case of the QRV method is presented for a given regularization parameter β; in general, the total CPU time increases by a factor of $\Re$, where $\Re$ is the number of runs required to determine the optimal regularization parameter β^*. The numerical solution of the Stokes problem (by the conjugate gradient method) is the most time consuming calculation: it takes about 180 s to reach a high accuracy in computations of the velocity potential. The reduction in the CPU time for the QRV method is attained by employing the velocity potential computed at β_i as an initial guess function for the conjugate gradient method to compute the vector potential at β_{i+1}. An application of the VAR method requires computing the Stokes problem twice to determine the 'advected' and 'true' velocities (Ismail-Zadeh et al. 2004). The CPU time required to compute the backward heat problem using the TVD solver (e.g., Ismail-Zadeh and Tackley 2010; chapter 7.9) is about 3 s in the case of the QRV method and 2.5 s in the case of the BAD method. For the VAR case, the CPU time required to solve the direct and adjoint heat problems by the semi-Lagrangian method (e.g., Ismail-Zadeh and Tackley 2010; chapter 7.8) is $1.5 \times n$, where n is the number of iterations in the gradient method used to minimize the cost functional (Eqs. 3.5 and 3.6).

Data-driven numerical modelling is useful tool for improving our understanding of the thermal and dynamic evolution of the Earth's structures. We have presented the BAD, VAR and QRV methods for data assimilation and their realizations with the aim to restore the evolution of the thermal structures. The VAR and QRV methods have been compared to the BAD method. It is shown that the BAD method can be employed only in models of advection-dominated mantle flow, that is, in the regions where the Rayleigh number is high enough (e.g., >10^7), whereas the VAR and QRV methods are suitable for the use in models of conduction-dominated flow (lower Rayleigh numbers). The VAR method provides a higher accuracy of model restoration compared to the QRV method, meanwhile the latter method can be applied to assimilate both smooth and non-smooth data. Depending on a geodynamic problem one of the three methods can be employed in data-driven modelling.

References

Ismail-Zadeh A, Tackley P (2010) Computational methods for geodynamics. Cambridge University Press, Cambridge

Ismail-Zadeh AT, Talbot CJ, Volozh YA (2001) Dynamic restoration of profiles across diapiric salt structures: numerical approach and its applications. Tectonophysics 337:21–36

Ismail-Zadeh A, Schubert G, Tsepelev I, Korotkii A (2004) Inverse problem of thermal convection: numerical approach and application to mantle plume restoration. Phys Earth Planet Inter 145:99–114

Ismail-Zadeh A, Schubert G, Tsepelev I, Korotkii A (2006) Three-dimensional forward and backward numerical modeling of mantle plume evolution: effects of thermal diffusion. J Geophys Res 111:B06401. doi:10.1029/2005JB003782

Ismail-Zadeh A, Korotkii A, Schubert G, Tsepelev I (2007) Quasi-reversibility method for data assimilation in models of mantle dynamics. Geophys J Int 170:1381–1398

Lattes R, Lions JL (1969) The method of quasi-reversibility: applications to partial differential equations. Elsevier, New York

Samarskii AA, Vabishchevich PN (2007) Numerical methods for solving inverse problems of mathematical physics. De Groyter, Berlin

Tikhonov AN, Arsenin VY (1977) Solution of ill-posed problems. Winston, Washington, DC

Tikhonov AN, Samarskii AA (1990) Equations of mathematical physics. Dover Publications, New York

Printed in the United States
By Bookmasters